BIOLOGY, HATCHERY AND CULTURE TECHNOLOGY OF TIGER PRAWN AND GIANT FRESHWATER PRAWN

Biology Hatchery and Culture Technology of Tiger Prawn and Giant Freshwater Prawn

Chiranjib Chakraborty
Arup Kumar Sadhu

2024

Daya Publishing House®
A Division of
Astral International Pvt. Ltd.
New Delhi – 110 002

First Published, 2001

Reprinted, 2024

ISBN 9788170352310

Published by : **Daya Publishing House®**
A Division of
Astral International Pvt. Ltd.
– ISO 9001:2015 Certified Company –
4736/23, Ansari Road, Darya Ganj
New Delhi-110 002
Ph. 011-43549197, 23278134
E-mail: info@astralint.com
Website: www.astralint.com

Foreword

Prawn farming is emerging as an important area of value addition and diversification in the aquaculture practices in the country in the recent years. Contributing to the valuable export earnings, tiger prawn, *Penaeus monodon* has become the most cultured species in brackishwater aquaculture, while giant prawn, *Macrobrachium rosenbergii* is being considered for culture across the country. In the context of utilising of brackishwater in the inland States of Punjab, Haryana, Uttar Pradesh and Rajasthan, prawn farming, both tiger and freshwater prawn, assumes a great significance.

With increasing entrepreneurship interest in prawn farming, it has become necessasry to provide technical material in different forms like books, bulletins, etc. It is appropriate that the present publication on "Biology, hatchery and culture technology of tiger prawn and giant freshwater prawn" by Dr. Chiranjib Chakraborty and Dr. Arup Kumar Sadhu is being presented at this time. I am sure, the publication would be of benefit to all those concerned with prawn farming in the country.

— **Dr. A.K. Ayyapan**
Director
Central Institute of Freshwater
Aquaculture (CIFA)
Kausalyaganga
Bhubaneswar - 2
INDIA

Preface

In India, nescent aquaculture industry is evolving fast at a pace witnessed never before. This aquaplosion has necessitated the dissemination of comprehensive information of biological, management, engineering as well as economic aspect coupled with it.

This book is sincere effort to meet the need and demand of both – professionals and academicians in various spheres of aquaculture, as felt by the authors for the development of aquaculture in various states in India.

Aquaculture specially tiger prawn and giant freshwater prawn has tremendous scope in tropical countries. Millions of hectares of potentials freshwater as well as coastal land awaiting for its development for prawn aquaculture can provide daily bread for several million people besides uplifting the status of rural community as their country.

This book is an endeavour to generate disciplined manpower for the prawn farming as it includes almost all basic ideas and compact form of scattered information on the prawn system, prawn biology, design, construction and management of hatchery and culture and the associated economics.

In the preparation of manuscripts, help has been taken from various source materials, which have been duly acknowledged in the text. Every effort has been made to properly acknowledge the various source of information included in the script. Omissions if any are inadvertent.

Grateful thanks are due to Director and Mr. Anil K. Mittal, Daya Publishing House for publishing of this manuscript and Smt. Shelly Bose, Head, Dept. of English, Women's College, Burdwan for correcting this manuscript. Criticisms and suggestions are welcome from the readers for improvement of the text in its subsequent editions.

Christmas, 1999

Dr. Chiranjib Chakraborty
Dr. Anup Kumar Sadhu

Contents

Introduction

Water is life. Plentiful water is available on our planet, Earth. Globally it is said that 2/3rd of the earth's surface is covered with water of which 97.2 per cent is salinewater and 2.8 per cent freshwater. It is said that "Fish is the queen of water" and so in this light that investigation on water resources for fish production purposes is made. But now, the demand of shell–fish resources is more high than fin–fish resource to the professional and aquaculturist.

Prawns are the most valuable aquatic fin–fish resources commanding good domestic and export market. Their economic importance is pretty high as they are among the choicest food items.

Indian's prawns grounds are among the world's most productive ones noted both for their high yield and superior quality of species occurring in fairly high proportions, among them, the giant freshwater prawn and the tiger prawn are most important. These two prawns have become an attractive commodity of both large and smallscale aquaculture enterprises.

Out of the total 1.4 million ha of potential brackish water area available in our country, so far 90,000 ha only has been brought under culture of tiger prawn farming. In case of fresh water, areas available for culture based capture are estimated 2.25 million ha of ponds and tanks. 1.30 million ha of beels, jheels, 2.09 million ha lakes and reservoirs and 0.12 million km of canals and channels and 2.30 million ha of paddy fields, a portion of which is amenable to aquaculture.

Since the growing demand of the prawn, farmers have already started farming. In the near future, aquaplosion of these two species will be more and more to meet the increasing demand.

Chapter 1

General Biology of Prawns, Biology of *Penaeus monodon*, Biology of *Macrobrachium rosenbergii*

I. General Biology of Prawns

1. Exoskeletons and Body Parts

The body is distinctly divided into two parts – *Cephalothorax and abdomen*. Both these parts are covered by hard exoskeleton.

Cephalothorax – It is formed by the fusion of head and thorax. A continuous shield-like exoskeletal covering, called carapace, encloses the cephalothorax. On both the ventrolateral sides, the carapace hangs freely over the gill-chamber as gill-cover which is commonly called as branchiostegite. On the ventral side the branchiostegite bears a thin membrane, branchiostegeal membrane.

Ventrally, the carapace covered by several hard sterna plates which carry paired appendages are specialised for different function. Following structures are present on the cephalothoracic region.

2. Cephalothoracic Appendages to Illustrate their Variety in Structure and Function

Segment	Function
1. Rostrum	Defensive
2. Compound eye	Sensory
3. 1st antenna or Antennule	Sensory (Base bears – Statocyst as a outer feeler carries olfactory setae, probably for determining smell)

Segment	Function
4. 2nd antenna– Mandible	Scale Serves as a balancer. Flagellum carries numerous tactile setae. Base bears green gland – excretory organ. Crushes food
5. First maxilla	Pass food to mouth.
6. Second maxilla	Pass food to mouth and drive water out or respiratory cavity.
7. 1st maxilleped	Mince food and pass it to mouth.
8. 2nd maxilliped	
9. 3rd maxilliped	
10. Walking leg (Chelate)	Chelate legs – Locomotion, offence and defence and can seize small prey
11. Walking leg (Chelate)	
12. Walking leg (nonchelate)	Locomotion – Transfer of sperm in mating
13. Walking leg (nonchelate)	
14. Walking leg (nonchelate)	
15. Pleopod or swimmerets	Function in forward swimming in female. There are also a place for attachment of eggs
16. Pleopod or swimmerets	
17. Pleopod or swimmerets	
18. Pleopod or swimmerets	
19. Pleopod or swimmerets	
20. Uropods or enlarged swimmerets	Tail fan, changing direction and backward swimming.
21. Telson	

3. Abdomen

A laterally flattened abdomen made up of six segments and a posterior most pointed part is the telson. Each abdominal segment is bounded by a ring-like exoskeletal piece called the sclerite. Each Sclerite consists of a dorsal arch shaped tergum which suspends freely on the lateral sides as pleuron and a ventral plate–like sternum. Each abdominal segment carries ventrally a pair of appendages which are called pleopods or swimmerets. The last pair is modified to a enlarged swimmeret and known as uropods.

4. Internal Organs

4.1 Digestive Structure

The digestive structure is composed of *alimentary* canal and *digestive gland.*

4.1.1 ***Alimentary Canal:*** It consists of foregut, midgut and hindgut.

A. Foregut: *Mouth* – Present on the ventral side of cephalothorax between the 3rd and 4th segment. Mouth opening bounded anteriorly by labrum, posteriorly by two lobed labium and laterally by the mandible.

Buccal Cavity – Flattened chamber between mouth and oesophagus.

Oesophagus – Short and wide tube runs vertically from the buccal cavity and leads to the stomach.

Stomach – Occupies longitudinally most of the cephalothoracic cavity, divided into two parts – a larger anteriorly bag like cardiac stomach and much smaller posterior stomach.

B. *Midgut:* The midgut is a narrow, long and elongated tube, which starts from the pyloric stomach and runs along the median line upto sixth abdominal segment.

C. *Hindgut:* Leading from midgut to the anus. Its anterior swollen muscular part, called the rectum which opens to the exterior through the anus. It is a slit-like opening, present near the base of the telson.

4.1.2 ***Digestive Gland:*** Large, bilobed, orange-coloured organ which is known as hepatopancreas. It surrounds completely the pyloric stomach and the two large hepatopanceatic ducts open into the ventral chamber of pyloric stomach. This digestive gland serves as liver, pancreas of higher animals. It absorbs digestive food and can store it for future use.

4.2 Respiratory Structure

Three sets of respiratory organs – gills, epipodites branchiostegite lining are enclosed by the branchiostegite which is opened in all side except dorsally.

A. *Gills:* On each lateral side of the cephalothorax there are eight gills, each attached with the thoracic wall by a gill root. Seven of these eight gills are serially arranged, which the eight gill is concealed under the second gill. Each gill consists of a slender axis on which double row of leaf like gill plates are arranged.

B. *Epipodites:* Highly vascularised leaf-like membranous structures present one on the coxal segment of each maxilliped.

C. *Branchiostegite lining:* The richly vascularised membrane of the branchiostegite serves as respiratory surface.

4.3 Blood Circulatory Structure

The circulatory system consists of circulatory fluid – blood, heart, true blood vessels and haemocoelomic spaces.

A. *Blood*

a) The cellular part of the blood includes only amoeboid leucocytes.

b) The plasma contains a copper containing respiratory pigment haemocyanin in dissolved state.

B. *Heart*

a) Triangular organ and placed beneath the carapace and above the gonads.

b) The entire structure is enclosed with a haemocoelomic space called pericardial sinus the wall of which serves as pericardium.

c) The wall of the heart contain five pairs of slit like opening called ostia.

C. *Blood Vessels*

a) From the heart of prawn six large vessels – *single opthalmic artery, paired antennary arteries, paired hepatopancreatic arteries* and a *single midposterior artery* emerges from the posterior end.

b) Opthalmic artery emerges from the apex of the heart and runs anteriorly along the mid-dorsal line upto the base of the rostrum.

c) Each antennary artery originating from heart runs anteriorly along the sides of opthalmic artery and unites with both antennary arteries and middle opthalmic artery at the anterior end. Each antennary artery, the pericardial gastric, mandibular, ventral and dorsal branches comes out from each side.

d) From the posterior end of the heart, the hepatopancreatic artery originates from each side and supplies the hepatopancreatic region.

e) Midposterior artery originating from posterior end of the heart divides into two main branches i.e. supra–intestinal and sternal.

D. Haemocoelomic Spaces

All the arteries ultimately break up into finer branches and open within the haemocoelomic spaces. Small haemocoelomic spaces called lacunae and large one is sinus. Blood after flowing through small haemocoelomic spaces is collected in a pair of common elongated space called ventral sinus. The two ventral sinuses are interconnected by several small slender channels.

4.4 Excretory Structure

Green gland is the main excretory organ which are remained within the bases of each second antenna. Each gland is composed of four regions, a blind–ending sac, green tube known as labyrinth, a long white nephridial canal and a thin–walled bladder or renal sac.

4.5 Reproductive Structure

Bisexual, gonads are of different shapes and both occupy similar position i.e. above the hepatopancreas and beneath the hear.

4.5.1 Male Reproductive Structure

a) Tests are paired, elongated and whitish in colour, united anteriorly and posterior end are free.

b) From the posterior end of each testis, a long much coiled tubular duct, called vasdeferens originates.

c) Each vasdeferens runs obliquely downwards and backwards between the thoracic wall towards the fifth walking legs.

d) Vasdeferens near the base of the fifth walking legs is swollen as seminal vesicle where sperms are temporarily stored and transformed into small packets called spermatophores.

e) Seminal vesicle opens at the base of fifth walking legs as male gonopore, which is guarded by a small circular lid.

4.5.2 Female Reproductive Structure

a) Paired, white ovary present on the same position which are united at their both end.

b) From the outer border and near the middle of each ovary a short and wide oviduct originates and runs straight downwards to the third walking legs.

c) On the inner side of each third walking leg, the oviduct opens outside as a female gonopore.

4.6 Nervous Structure

Nervous structure consists of central nervous, peripheral and sympathetic nervous structure.

4.6.1 Central Nervous Structure

It runs from anterior to posterior end and contains following structures:

a) A pair of supra–oesophageal ganglia placed dorsally and near the base of rostrum and are known as brain.

b) In thoracic region, a large ventral elongated mass of ganglion are known as thoracic ganglionic mass.

c) A paired cords, each of which begins from the supra–oesophageal ganglion of one side and runs posteriorly along the ventrolateral wall of the cephalothoracic cavity and ultimately unite with thoracic ganglion.

d) From the posterior end of thoracic ganglion, two separate cords appear posteriorly and passes upto the posterior most region as a ventral nerve cord. It bears a ganglion in each segment. The last abdominal ganglion are large and known as stellate ganglion.

4.6.2 Peripheral Nervous Structure

a) Optic Nerve – From anterior side of each lobe of brain, an optic nerve enters within the eye.

b) Antennular Nerve – It originates from the same regions and innervates in the first antenna.

c) Antennary Nerve – From the Posterior side, the antennary nerve originates and supplies to the second antenna.

d) Cephalothoracic Nerve – Eleven pairs of thoracic nerves

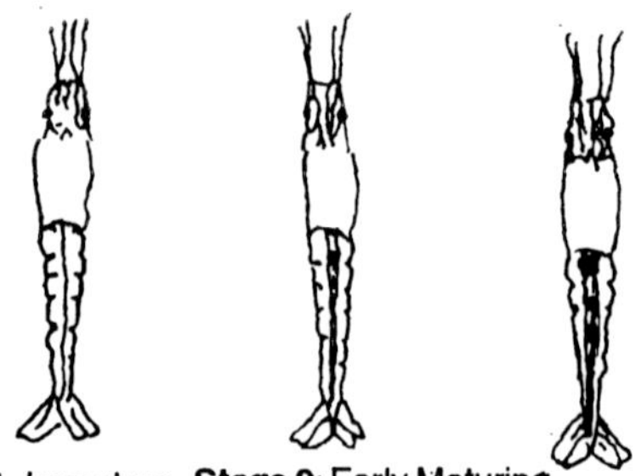

Stage 1: Immature **Stage 2**: Early Maturing **Stage 3**: Late Maturing

Stage 4: Mature

Fig: Ovarian Development of *P.monodon.*

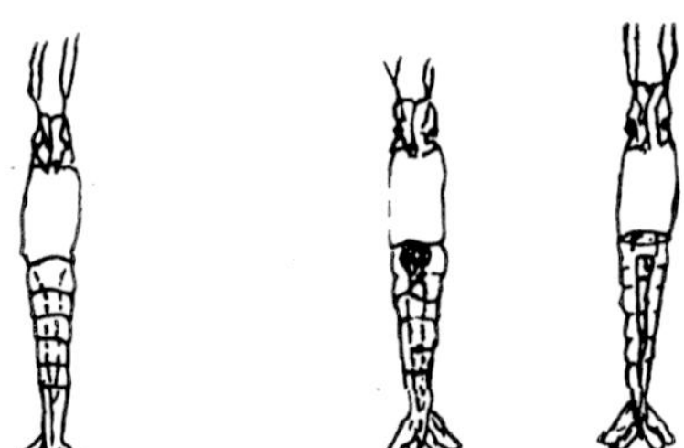

Complete Spawning Partial Spawning

Fig: Post Spawning Appearance of *P.monodon.*

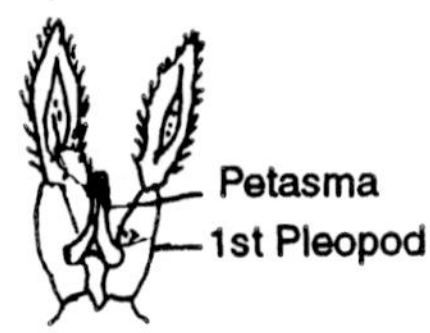

Fig: Male External Genitalia of *P.monodon.*

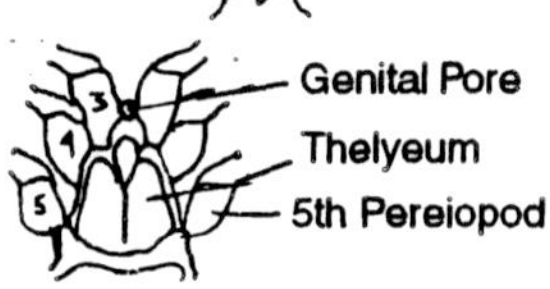

Fig: Female External Genitalia of *P.monodon.*

Photo 1: Lateral view of *Penaeus monodon*.

Photo 2: Dorsal view of *Penaeus monodon*.

emerge from the thoracic ganglion to supply different muscles and appendages in that region.

e) Abdominal Nerve – Two pairs of pairs of peripheral nerves originate from each abdominal ganglion are given to supply muscles and appendages in that region.

4.6.3 Sympathetic Nervous Structure

A few minute ganglia and slender nerves are present over the cardiac stomach to supply the involuntary parts of the body.

II. Biology of *Penaeus monodon*

1. Distribution

Penaeus monodon occurs in the brackish waters, estuaries and in the inshore and deeper waters of the east and west coasts, but its availability is much lesser than other marine prawns like *P. indicus*. *P. monodon* completes its life cycle in two environments, the marine and estuarine environment. It is uryhaline species with high tolerance to fluctuations of salinity.

Its general distribution is along the entire east coast of Africa, madagascar, Mauritius, Pakistan, east and west coasts of India, Sri Lanka, Andaman Islands, Singapore, Malaysia, Philippines, Southern Japan, Celebes and Nearby Islands, Taiwan and Queensland and New South Wales in Australia.

2. Systematic Position

Phylum	:	Arthopoda
Class	:	Crustacea
Subclass	:	Malacostraca
Order	:	Decapoda
Suborder	:	Natantia
Family	:	Penaedae
Genus	:	Penaeus
Species	:	*monodon* (Fabricius)

3. Species Identity

a) Rostrum long, sigmoid with 7-8 dorsal teeth and 2-3 ventral teeth.

b) Esopodites of the fifth walking legs absent.

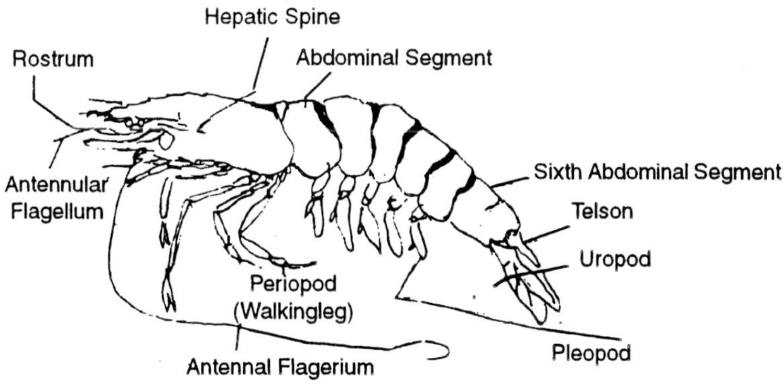

Fig: External Feature of a *P. monodon.*

c) Petasma with median anterior lobe small and not reaching the length of the lateral lobe.

d) Seminal receptacle in thelycum is circular.

e) General colouration dark blue to almost black with dark bands across carapace and abdomen, pleopods and uropods tipped blue.

4. Food and Feeding Habits

Crustaceans form the main food item comprising the Harpacticoid copepods and parts of brachyureans, other item of food are molluscs, parts of fish polychates and vegetable matters.

The food of the fry in the cultivating ponds is 'lab lab', blue green algae and a mixture of minute plant and animal organisms.

The larvae during zoeal stages feeds on algae, mainly diatoms like *Chaetoceros* sp. Mysis stage onwards feeds on minute zooplankton like *Artemia* sp. or phytoplankton. Post larvae are carnivorous and prefer to hunt the zooplankton.

5. Growth

This is a very fast growing prawn. In the Philippine culture ponds fry measuring 15.3mm in length have been observed to attain an average increases in length to 94.7mm, 149.9mm and 229.8mm at the end of three months, six months and one year respectively, with average weight of 6.88gm, 22.3gm and 95.1gm. It has also been shown that in those ponds the growth was even faster in some cases the largest attaining length of 250mm weighing upto 120gm at the end of the year.

6. Male and Female Identification

6.1 Male

The male have external reproductive organ called petasma which is found in the first abdominal segment.

6.2 Female

The female can be identified by the presence of the external reproductive organ called thelycum which is situated at the base of 4th and 5th pair of walking legs.

7. Maturity

The species grows to maturity when their sizes are around 200mm (weighing about 65gms) in male and 240mm (weighing about 90gm) in female.

The key indicator is the broadening of the ovary in the first abdominal segment which gives a characteristic diamond shape of ovary as well as the diamond pattern in the first segment.

8. Induced Maturation by Eye Stalk Ablation

In time of hatchery operation, mainly females are induced for maturation by unilateral eyestalk ablation. Ablation is carried out by making an incision across the eyeball with a razor blade, then squeezing out all the contents through the incision by applying firm pressure starting at the base of eyestalk and working upwards towards the slit in the eyeball.

9. Fecundity

Fecundity ranges from 80-200 thousand eggs depending upon size of the female. Fecundity is more in the wild than in pond reared brood stock on account of different growth rate. In nature, 3 or 4 spawning may occur in a year.

10. Life Cycle

P. monodon dwells in the oceanic waters, it has a coastal estuarine phase in its life cycle. The adults mature, mate and release eggs in the deep oceanic waters. The hatched out larvae dwell in the surface waters of the ocean till they reach post larval

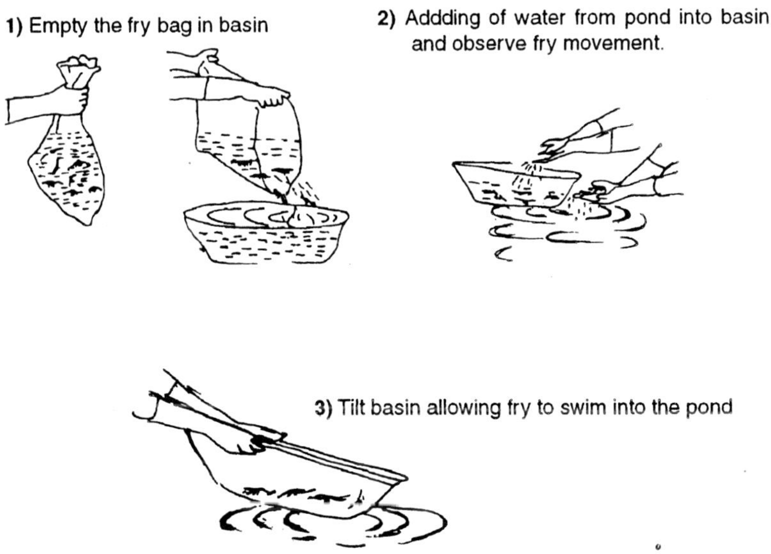

Fig: Acclimatization of Prawn Fry for Culture

stage. The post larvae will drift towards coastal water (Brakish water) areas which serves as nursery ground for them. As these grow to sub-adult stage, they start migrating towards deeper oceanic waters, where they mature and reproduce.

The larval phase of *P. monodon* life cycle are spent in oceanic waters with a salinity of 28 to 35 ppt. The life cycle completes through nauplius, zoea, mysis, post larva and adult.

10.1 Nauplius

Nauplius are pyriform in shape, dark brown in colour, opaque and swim in short jerks by means of three pairs of appendages – first antennae, second antennae and mandibles. This stage has six substages, (N_I to N_{VI}).

10.2 Zoea

This is the first feeding stage. Its body is elongated and carapace, thorax, abdomen are distinct. Antennae, mandibles, maxillae and maxillipede are also distinct and are useful in filter feeding and swimming (7 to 8 pairs of appendages). It has three substages (Z_I to Z_{III}).

10.3 Mysis

Mysis has well developed abdomen and carapace. The five peraeopods have undergone considerable enlargement. The exopods serves swimming organs (13 pairs of appendages). Mysis keeps its head down and swims. It has three substages (MI to MIII).

Fig: Larval Stages of *P.monodon.*

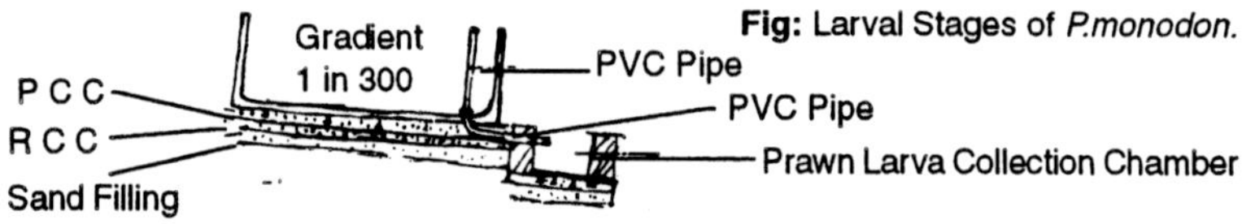

Fig: Cross Section of AB

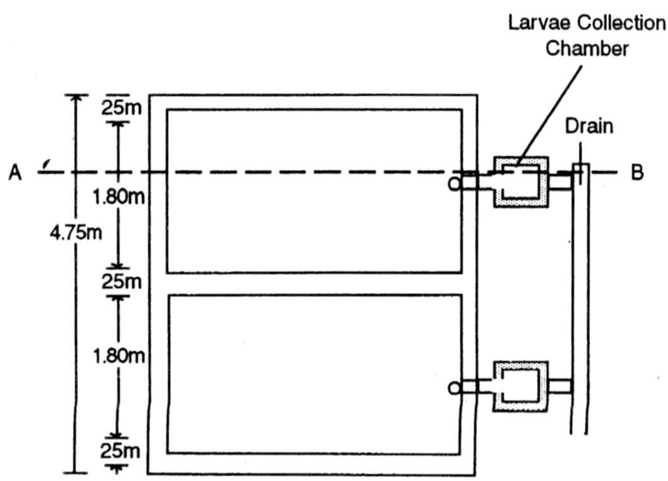

Fig: Plan of One Set Larval Rearing Tanks

10.4 Post Larva

Fully developed post larvae resembles the adult in its shape. It has well developed pleopodes, which serve as main swimming organ (19 pairs of appendages). It swims horizontal like to adult. The post larva are aged in days. After its first appearance, calling PL1 on its first day and PL2 on its second day and so on. These stages tolerates lower salinities.

Table : Life Span and Identification Features of Larval Stages of *P. mondon*

Stage	Life Span	Identification Features
Hatching	12 – 15 hrs. after spawning	Furcal Spines 1+1, the setae on appendages are not plumose
NI	5 – 6 hrs.	
NII	5 – 6 hrs.	Furcal spines 1+1, the setae on appendages are plumose (well developed)
NIII	5 – 6 hrs.	Furcal spines 3+3
NIV	5 – 6 hrs.	Furcal spines 4+4
NV	5 – 6 hrs.	Furcal spines 5+5
NVI	15 – 20 hrs.	Furcal spines 6+6 then 6+7
ZI	1½ – 2 day	No eye stalks.
ZII	1 day	Stalked eyes.
ZIII	1½ – 2 day	A pair of biramous uropods and dorsomedian spines on abdomen.
MI	1½ – 2 day	Rudiments of pleopods (Mysis swims its head down).
MII	1 day	Unsegmented pleopods with terminal setae.
MIII	1½ – 2 day	Segmented pleopods with terminal setae.
PL1	1 day	Segmente pleopods with setae will over (post larvae swims horizontally)

III. Biology of *Macrobrachium rosenbergii*

1. Distribution

M. rosenbergii is found in tropical and subtropical areas of Indo-Pacific region including India. For culturing purpose Japan,

Singapore, Australia, Hawaii and many countries of Latin America obtain directly or indirectly their original stock from Malayasia. But Bangladesh, Burma, India, Indonesia, Kampuchea, Malayasia, Philippines, Srilanka, Thailand and Vietnam – all have their own natural stock. In India, the species is distributed in the lower stretches of most of the river systems of both the east and west coast region. It forms considerable fishery in Hoogly, Godavari and Narmada river systems, Kollery Lake and Kerala back waters.

2. Systematic Position

Phylum	:	Arthropoda
Class	:	Crustacea
Subclass	:	Malacostraca
Order	:	Decapoda
Suborder	:	Natantia
Infraorder	:	Caridea
Super Family	:	Palaemonidae

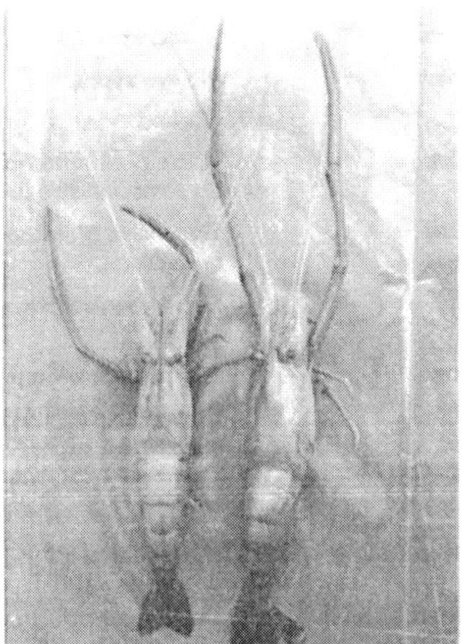

Photo 3: Dorsal view of female and male of *Macrobrachium rosenbergii.*

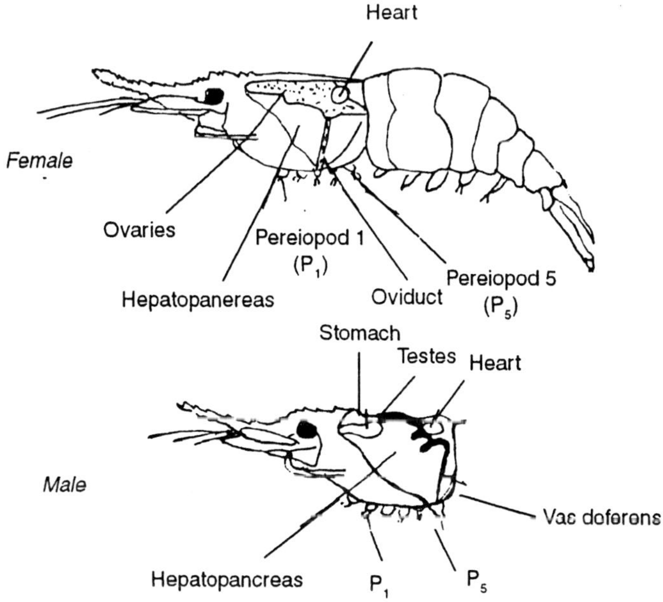

Fig: Internal Reproductive Structure of *Macrobrachium rosenbergii.*

Family	:	Palaemonidae
Sub Family	:	Palaemonidae
Genus	:	*Macrobrachium*
Species	:	*rosenbergi* (de man, 1879)

3. Species Identity

a) Rostrum is bent in the middle and upturned distally. It is armed with 12 to 13 teeth on dorsal, and 11 to 13 on ventral margin. The first two or three dorsal rostrum teeth are on the carapace.

b) The length of the rostrum is equal or more than the length of the carapace.

c) Branchiostegal spine absent, hepatic spine present.

4. Food and Feeding Habits

Giant freshwater prawns are omnivorous and under natural conditions feed on aquatic products – vegetable waste, zooplankton and microscopic fauna living on the bottom i.e. insect larvae, small worms and microorganisms. It accepts a variety of food items as a supplementary feed ranging from grains, worms, flesh pieces of mollies, crustaceans and fish and cooked egg pieces.

The species is nocturnal in habit and most of its life activities especially moulting and hatching takes place during night times. During daytime it is sluggish and tries to hide or bury in the bottom. There is a tendency to establish territory and protect the same in the adults.

It locates its feed mostly by touch with feeders. The first pair of chlipod is the chief organs of food capture and is assisted by the second pair.

5. Growth

Like other crustaceans that have a rigid exoskeletoa, growth is essentially as discontinuous process achieved through a succession of moults separated by intermolt periods. Almost all growth occur incrementally, immediately after a moult and before hardening of the new integument. After the principal growth phase, growth still occurs due to the flexibility of the membranes that join the plates forming the rigid skeleton but at a comparatively much slower rate. The intermoult period gradually increases with age and in the fully grown adults, the growth stops leading to senility and death. Fully grown adults have greenish mat of periphyton growing on the cephalothorax, which indicated that the growth has stopped.

6. Male and Female Identification

6.1 Male

a) The cephalothorax is bigger in size and the abdominal space is narrower.

b) The 2nd of the five pairs of walking legs are larger and thicker than the others and ends in a more pronounced claw.

c) The genital pores are between the bases of 5th walking leg.

6.2 Female

a) The cephalothorax of the mature female and its 2nd pair of walking legs are much slender than the adult male.

b) The first three abdominal pleura are elongated ventrally to make space for holding eggs, popularly known as 'brood-pouch'.

c) After prespawning moult oviparous setae are developed on the inner margin of the bases of the first four pairs of pleopods which served to anchor the eggs to the pleopods for incubation.

d) The genital pores are located at the bases of the 3rd walking legs.

7. Maturity

The species grows to maturity when their sizes are around 150mm (30–35gm) in females and 175mm (35gm) in males.

Mature oviparous females are easily identifies because the ovaries can be seen as large coloured masses occupying a large portion of the dorsal an lateral parts of the cephalothorax.

8. Breeding

During the breeding season before mating the females go through a prespawning moult. Within 3 – 6 hours the mature male hold the soft bodies female between its chelate legs and deposits sperms as a gelatinous mass near the genital pores of the female located at the base of the 3rd pair of thoracic legs. The female lays eggs 6 to 20 hours after mating, and they get fertilized externally. The eggs are deposited into the brood chambers under the abdomen between the pleopods. The fertilized eggs are held in bundles by an extremely this elastic membrane, considered to be secreted by the oviparous setae of the first four pairs of pleopods. The adult egg carrying female at this stage are known as berried females

9. Fecundity

Fecundity ranges from 40 – 150 thousand eggs depending upon size of the female. A medium prawn of 180mm long and 60gm weight produces 60,000 eggs. Fecundity is more in the wild than in pond reared brood stock on account of differential growth rate. In tanks mature females spawned twice in five months. In nature also 3 or 4 spawning may occur in a year.

10. Incubation

The females continuously incubate the eggs and provide sufficient aeration for the developing eggs by constantly vigorous movement of the pleopods. Incubation of the fertilized eggs takes place fro a period of 18 to 21 day depending on the temperature. (19 – 20 days at 25 – 32⁰C). Dead eggs and foreign matter is carefully removed by the first pair of thoracic legs. From about 12th day onwards, a light grey colour slowly develops in place of orange colour. When the eggs are about to hatch, the developing eye can be seen as a black spot. Incubation lasts 15 to 20 days depending on the temperature.

11. Life Cycle

The life cycle of the species consists of eggs, larvae (zoea), post larva, a juvenile, adult, Maturity, mating, egg–laying and part of incubation takes place in freshwater habitat and the berried females migrates to suitable brackish water environment for hatching, growth of larvae, post larvae and juveniles. The juvenile ascends into rivers, canals etc. to grow into adults.

11.1 Embryonic Development

Day	Development
1st day	First division takes place at about four hours after fertilization and subsequent division takes place at 90 minutes intervals. Cleavage is completed in about 24 hours.
2nd day	Germinal disc appears on the ventral side of embryo. Gastrulation takes place followed by the formation of blastopore. A v–shaped germinal fends is differentiated from the blastoderm on the ventral surface and assuming the position of the developing embryo.
3rd day	The rudiments of embryonic regions are formed and the embryo develops into the embryonic nauplius stage.
4th day	Development of appendages of the nauplius takes place.
5th day	Post nauplius segment begin to be formed followed by the appearance of rudiments of other cephalic appendages.
6th day	The caudal papilla appears with rudiments of thoracic legs and telson.
7th day	The optic rudiments develop and rudiments of appendages and caudal papilla elongate to becomes the telson.
8th & 9th day	Eye pigment develops, the carapace rudiment is formed.
10th & 11th day	The heart begins to beat and gut is formed.

Contd. ....

Day	Development
12th to 16th day	The embryo lies parallel to the long axis of the egg and the larva steadily grows.
17th day	Appendages elongate and organs are fully formed.
19th to 20th day	The larva hatches out.

11.2 Hatching Behaviour

The hatching is accomplished with increased internal pressure due to stretching of body and continuous vibration (slowly) of mouth parts of larvae. After an hour, thoracic appendages vibrate vigorously but intermittently for about 10 minutes initially. The body stretches further and telson thrust out first followed by head. Immediately the larvae come down to the bottom of the tank and stay for some time inactively (called prezoea stage). Gradually, the larvae gains activity, comes up and leads a planktonic life.

11.3 Larval Development

Larvae pass through eleven moults but only eight morphological stages can be recognized distinctly. The larval cycle of the species lasting from 30 – 45 days depending upon temperature, water quality and food. The growth rate is fairly even during the first three larval stages, but later becomes irregular. Identification characters, sizes and number of days (at 28^{0}C for each stage) are given below:

Stages	Age (No. of days)	Body Length	Important Character
I	0 (1–2)	1.92 mm	Sessile eyes
II	2 (2–3)	1.99 mm	Stalked eyes
III	4 (3–5)	2.14 mm	Uropods appear
IV	7 (5–9)	2.50 mm	Two dorsal teeth on rostrum
V	10 (9–12)	2.80 mm	Telson narrower & elongated
VI	14 (12–18)	3.75 mm	Pleopod buds appear
VII	17 (15–20)	4.06 mm	Pleopod biramous
VIII	20 (18–22)	4.68 mm	Pleopods with setae
IX	24 (21–29)	6.07 mm	Endopods of pleopods with appendix interna

Contd......

Stages	Age (No. of days)	Body Length	Important Character
X	28 (25–34)	7.05 mm	Three or four dorsal teeth on rostrum
XI	31 (28–37)	7.73 mm	Teeth on half the upper dorsal margin
P.L.	36 (36–43)	7.69 mm	Teeth of upper and lower margins of rostrum

11.4 Food Habits and Behaviour of Larva

All larval stage are omnivorous in feeding habits, subsisting on zoo–plankton and phytoplankton. They can also live on organic food like bit of flesh, prawn, molluscs, etc. Maxillipeds and thoracic legs are used for trapping food particles while swimming, preferring living and suspended food particles.

These are active swimmers and planktonic in habit. They are attracted to light, but not very strong illumination. They swim tail up, head down and ventral side upward at an oblique position. Upto 5th stage (about 10 days), there is schooling habit in the larvae as they tend to be close together and move in swarms. From 6th stage onwards the larvae gradually tend to disperse. They spend most of their time at the surface and in mid–column but settle down to bottom during moulting time.

11.5 Post Larvae

On completion of their larval life, the larvae metamorphoses into post larvae. Post–larva settles to the bottom and becomes benthic in habit. They are virtually miniature adult since their morphology is the same except for the under development of body parts like setae, spine, teeth etc.

Post larvae exhibit good tolerance to a wide range of salinity, which is a characteristic of freshwater prawn. Post larvae begin to migrate upstream into fish water conditions within one or two weeks after metamorphosis and are soon able to swim against rapidly flowing currents where they spend their adult lives.

11.6 Juveniles

Metamorphosis of post larva into juvenile is characterized by a radical change in behaviour and appearance. They cease their pelagic life, sink down and settle on the bottom as crawlers and cling to vegetation and submerged objects. They feed on small worms, crustaceans, insect larvae as well as on organic matter of animal and plant origin. As the juveniles grow into sub–adults (about 70 to 80mm), these bonds starts disappearing. In nature juveniles remain in brackish water for one or two weeks, and then slowly migrate upstream into water of less salinity where they grow, mature and reproduce.

Chapter 2

Hatchery Technology of Tiger Prawn *Penaeus monodon*

1. Introduction

India has 1.42 million hectare of brakish water area suitable for fish and shrimp culture of which some area is exclusively under shrimp farming. A rapid development appears in India in shrimp culture field during recent years. For this rapidly changing scenario, requirement of shrimp seed has become many folds. Natural collection of shrimp seed is also very limited – which keeps a big gap between the demand and the supply of seed. This leads to establish commercial hatcheries for production of more seed under controlled environment. The Andhra Pradesh shrimp seed production supply and research centre (TASPARC) at Mangamaripeta near Visakhapatnam and the Orissa shrimp seed production supply and research centre (OSPARC) at Gopalpur in Orissa state were the first two commercial shrimp hatcheries, started functioning in India since 1989. By the 2000, in India, there about 100 hatcheries producing seeds in mass scale. West Bengal, is also producing shrimp seed in commercial basis by four hatcheries at Digha and on hatchery at Dadanpatrabar.

For accelerating the tempo and to bring under shrimp farming at least 10 per cent of the area to achieve 0.1 million farmed shrimp production target, annually 40.000 million shrimp seed will be required. (Gopalakrishna, 1988). In fact, commercial hatcheries in the process of being established now in various maritime states to produce annually 2,000 million seed. So, more hatcheries establishment are needed.

2. Site Selection

The location of a shrimp hatchery should satisfy following requirements:

2.1 Sea Water Supply

The seawater used in hatchery should be clean, clear and relatively free from slit. The water geniality should be good with minimum fluctuation in salinity all year round. Suitable sites are usually found near sandy or rocky shores. Sites, which are not suitable for hatchery, include areas, which are heavily influenced by rain or turbulence. River mouths should be avoided as abrupt salinity change occurs after a heavy rainfall. An added advantage of having a site on rocky shores is that good quality seawater is relatively near the shore line. This reduces the cost piping installation and pumping. The hatchery site should also be fr from any inland water discharges containing agricultural or industrial wastes. The seawater with temperature 28°C to 32°C, salinity 28–34 ppt, pH 7.7–8.2 and suspended solids below 20 ppm, is required.

2.2 Accessibility

Ideally, a hatchery site should be selected in areas where there are active shrimp farming operations so that the seed produced can be easily transported and distributed to the grow out ponds. The site chosen for a hatchery must have an easy access to communication and transportation channels.

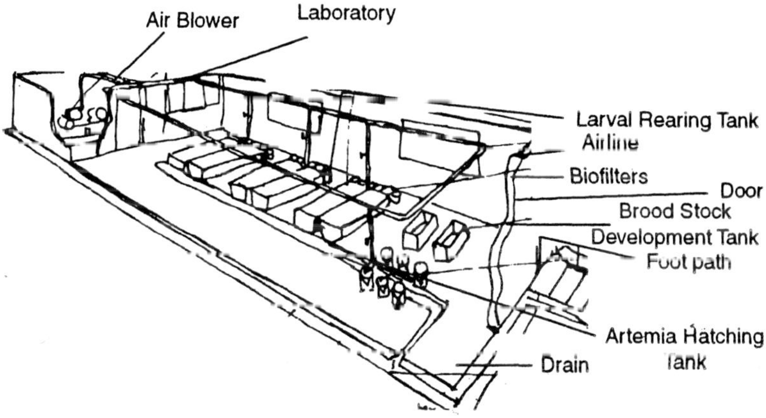

Fig: Layout of Giant Fresh Water Prawn Hatchery

2.3 Topography

The ideal site should be spacious, situated on flat to gently sloping grounds, well drained and not susceptible of floods, strong wave and tidal actions. It should be on compact soil and accessible by paved road

2.4 Availability of Power Sources

A shrimp hatchery cannot be operated without electricity. Electricity is essential to provide the necessary power to run equipment and other life support systems of the hatchery. Hence, the site must have a reliable source of power. Installation of a standby generator is absolutely necessary especially in areas where there are frequent and/or lengthy power failures and fluctuation or without power supply.

2.5 Other Requirements

The stable quality of fresh water is essential. There must be a source of broodstock of the farm for brood stock development nearer to the hatchery. Availability of technical hands and skilled labours are also needed.

It is also advisable to pay attention to land values early in the site selection phase to ensure that the site is available for purchase or lease and at a price consistent with the project budget.

3. Classification of Hatcheries

The hatcheries are classified into four categories:

3.1 Backyard Hatcheries

These hatcheries can produce 1–2 million seeds per year.

3.2 Small Scale Hatcheries

These hatcheries can produce upto10 million seeds per year.

3.3 Medium–Scale Hatcheries

These produce 10 to 40 million seeds per year.

3.4 Large Scale Hatcheries

The hatcheries, which produce seeds more than 40 million per year called large–scale hatchery.

4. Hatchery Design

Hudinaga and Kittsaka (1967) pioneered the mass seed production technique in Japan using 50–200 tones capacity cement tanks as spawning – cum – larval rearing tanks. The inorganic fertilizers were added into the tank periodically to produce algae (diatoms) for feeding the shrimp larvae.

The recent hatchery tanks used for spawning and larval rearing in Japan are large, ranging from 100 to 2000 ton capacity. They are suitable for spawning a large number of spawners at a time and for rearing the resultant post larvae by what is referred as the 'community culture method'. The indoor tanks of concrete or fiberglass are commonly use for spawning there. Cylindroconical, 150 lit capacity fiberglass tanks have been found to be very convenient (AQUACOP, 1983).

Mock and Neal (1974) developed the small tank hatchery system in USA, known as Galveston system. Cylindrical, funnel shaped, small 2 ton capacity tanks are used fro larval rearing in this method. In there, different size tanks are used for different functions like spawning, algal culture, *Artemia* feeding and rearing post larval rearing and maturation.

Kungvankij (1982) described a system, which combines the above systems. It includes spawning tanks with capacities of 1000–3000 lit capacity and nursery tanks with a capacity of 30–100 ton for rearing post larvae to the PL30 stage. This system is reported to maximized tank utilization in spawning and larval rearing of species like *P. mondon*.

Liao (1985) referred to a recently developed 'ladder system' hatchery in Taiwan, consisting of four interconnected tanks built on sloping ground, with the algal culture tank at the top, followed below by the rearing tank for mysis to the PL 1–5 stages and finally a larger tank for post–larvae, all built on below the other, with the descending water levels.

The designs have been further modified and adapted commercially depending upon the regional requirements.

The major requirements of a shrimp hatchery of a state of the modern art are following:

4.1 Sea Water Intake and Filtration Unit

Generally, pure seawater is drawn from the sea by a pumping unit through filter. The filter (bed) is made up of different layers of sand, pebbles and granite throughout which the water is drawn into a PVC pipes having slit opening. This is called slow sand filter. The rapid sand filters are also available in the market now a days. Here, the back wash process is easily done, so, this rapid sand filter works well. The water is pumped up into chlorination unit and then dechlorination unit. These tanks are of R.C.C. from there, the water is pumped into over head tank (OHT) which is also made up of R.C.C. and located at an elevated level adequate enough to flow the water to the hatchery by gravity.

4.2 Aeration Unit

Oil free air compressor is required in the hatchery to ensure oxygen supply for maturation rearing and related system. A pair of positive displacement blowers are required for a 10 million seed hatchery. The air supply grid with P.V.C. pipes is connected for all the facilities.

4.3 Power Supply Unit

The power requirement varies depending upon the capacity and type of hatchery. A 10 million/annum capacity hatchery needs around 60kw power supply. A stand by generator to combat power failure with the AC mains is needed.

4.4 Brood Stock Development and Maintenance Unit

To produce brood stock, near the ponds of 0.25 to 0.50 hectare area are constructed near the hatchery. Though wild broods are used now. But brood stock unit is very much required from the unavailability of wild brooders.

4.5 Maturation Unit

The maturation tank used in hatchery is generally of 10–16 tones capacity, round or rectangular with oval shape inside. A 50 mm diameter P.V.C stand pike regulates the water flow during 200 per cent water exchange. The maturation unit should be sound proof. A photo–period of 12 h height: 12h dark is maintained in maturation unit. Diffused illumination for 12 hrs a day is provided by fluorescent lights of 40w each suspended at 1m above the tank. The total lighting stimulates like a deep sea environment for breeding.

4.6 Spawning and Hatching Unit

Spawning and hatching unit can be combined with the maturation shed, free from song. FRP or HDPE tanks or 250 lit capacity are generally used to spawn the mature female.

4.7 Larval Rearing Unit

The larval rearing section is one of the main area of activity. The tanks used are either of parabolic shape and cement construction of cylindro–conical or round flat bottom and made of FRP material. R.C.C. tanks may also be used. Capacity of tanks differs from 2 tones to 10 tones water. Each tank has one out let with maximum slope (1:500) drawn to wards the out let which enables to drain out even the last drop of bottom water. Aeration from air grid is provided in each tank and throughout water supply unit treated sea water is filled in larval rearing tank (LRT). The inner surface of the tank is painted with green apoxy paint.

4.8 Post Larva Rearing Unit

The hatcheries having indoor post larval rearing facility use cemented tanks either of rectangular flat bottom or parabolic 'U' shape of which the latter is more efficient. The capacity varies from 10 to 30 tonnes of water. The inner surface of the tank is coated with good apoxy resign painting of fiber coating for maintaining better hygienic condition. The post larval unit (PL) is also well connected with water and aeration system.

4.9 Algal Culture Unit

Algal section has got two wings–indoor algal section and outdoor algal section. Indoor section is completely air conditioned. As the space is indoor air condition algal unit is limited, 20 lit to 200 lit carbuoys are shifted to out door algal culture section. For mass culture 1 to 6 ton cemented tank or 200 lit FRP tanks may also used.

4.10 *Artemia* Hatching Unit

For hatching of *Artemia* cyst 200–500 lit capacity FRP cylindro–conical shaped tank having bottom drain facility with a side glass window is needed. Continuous aeration and 40w light fixed at 20cm–1mt above the tank are provided.

4.11 Counting and Packing Unit

For counting, packing and transportation of seeds a separate section is engaged. Now a days it is just an open shed.

All these together around 130sq meter ground area per million seed. A totally indoor rearing facility of the hatchery will need 50–70 sq. meter area per million built up space, in a hatchery shed protected by walls and roofed with asbestos sheet interspersed with translucent FRP sheet for lighting.

5. Seed Production and Hatchery Technique

The seed production in hatchery totally depends on effective operation and successful functioning. Successful controlled spawning and larval rearing of shrimps was initiated by Hudinga

Photo 4: *Artemia* packs from different company.

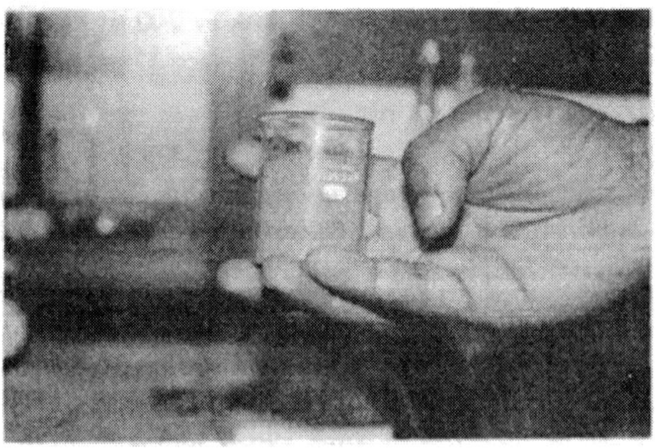

Photo 5: *Artemia* cyst in a 50ml beaker.

Photo 6: *Artemia* cyst hatching in a plastic basin by aeration.

(1942) with wild spawners of *Penaeus japonicus* catch from fishing grounds.

Following the early success of hatchery production of post larval of *P. japonicus* the considerable effort has been directed towards the controlled maturation, spawning and larval rearing of shrimps. The successful and developed methods for operation are the following:

5.1 Seawater Intake and Filtration

Quality sea water with 7.7–8.2 pH and 28–34 ppt salinity, free from agricultural, industrial and sewage pollution is drawn for the hatchery through water grid, fixed in sea coast. There are two storage tanks, one storage tank is used for chlorination and for dechlorine purpose and other one is used for storing the treated water. The sea water in the chlorination tank is treated with bleaching powder (5–10 ppm of chlorine) to kill all organisms causing diseases. The chlorination process takes 6 to 8 hours and after that the water is transferred through a vertical sand filter (filled with sand and pebbles) to remove the particulate and precipitates. The dechlorination is done by sodium thiosulphat (3–5ppm) depending upon the residual chloride level. DETA is added to act as chelating agent capable of removing chemical pollutants like trace metals. Then seawater is pumped to overhead tank, the water is then treated with uv ray to kill the micro organisms including bacteria and virus and this water is supplied to various units of the hatchery.

5.2 Aeration

Hatchery requires a continuous supply of aeration as to maintain a high level of DO constantly for brooders, larvae, post larvae, algae and *Artemia*. The air bubbles of the aerator lift the algal cell or food particles in suspension so that the larvae take the food particles easily. For any reason if aeration is stopped for 10–15 minutes, mass mortality occur in the larval rearing tanks specially for their high stocking density

5.3 Brood Stock Development and Maintenance

The number of breeders required for annual production of seed, is calculated the basis of average rate of nauplii production capacity per spawners and the rate of survival from the nauplii to PL20 stage under the environment condition of hatchery on an average in *P. mondon* a wild spawner yields 5 lakh nauplii and an artificially mature spawner yield nearabout 2 lakh naupalii. An average rate of survival to PL20 is around 30 per cent.

5.3.1 Brood Stock from Earthen Pond

To produce the brood stock, in earthen ponds, PL20 is stocked at the rate of 4 number per m^2. After 12 to 15 months they could be raised as a spawners using brackish water with required salinity in ponds. Full sea water salinity is required for the final stage in raising quality spawners as these shrimps attain maturity in nature only in offshore seawater. A captive spawner source from broodstock pond is advantageous for a hatchery for producing the seed throughout the year.

5.3.2 Brood Stock from Wild Catch

Brooders may be receive from wild catch. The brooders are transported by polythene pack early in the morning or night hours when the temperatures becomes low.

After transportation brooders are released into brood stock tank. A good number of aeration connection provided to the brood stock for sufficient oxygen to become stress free. From this stock, the better once are selected for maturation.

5.4 Brooder Maturation

The spawners brought from the source (pond or wild) are disinfected with formalin (50 ppm) for 30 minutes and quarantined for 1 to 2 days.

Prior to maturation, the good quality males are removed, measured and weighed and treated with antibiotics for 30 minutes. The treated specimens are transferred to maturation already filled with clean, filtered sea water passed through 10 micron cartridge filter. The female shrimps are then taken one by one, eyestalk is ablated. After ablation, the specimens are treated for 30 minutes with antibiotics to prevent infection. Then male and female 1:1 ration are released in the maturation tank.

With in 7–10 days the ovary develops. The overian development passes through four stages, which could be observed with the help of a light beam flashed from underneath its body in total darkness when the fourth stage of maturity is reached, the female is ready for spawning.

During the maturation, a vitamin and fatty oil rich feed like squid flesh or mussel meat, shrimp, goat or beef liver etc. are fed @ 10 per cent of the biomass of spawner shrimp. The maturation tank is daily cleaned to remove the waste very often. On alternate days the water is fully drained off and the prophylactic treatment is done.

5.5 Spawning and Hatching

Mother shrimps (gravid) are released in the spawning tanks after filling with treated seawater and aeration. Spawning takes place mostly in midnight hours. After spawning in the early morning, the mother shrimps are separated from the spawning tank to avoid damage to the eggs. Eggs are then lifted by siphoning and taken in the buckets for cleaning the eggs. After that through washing the eggs are shifted to hatching tanks (The hatching tanks are similar in size in spawning tank). Vigorous aeration is then provided. Eggs are hatched out into nauplii within 6–8 hours. The nauplii until metamorphose in zoea stage, they remain in hatching tank. The zoea are then shifted to larval rearing unit.

5.6 Larval Rearing

The larval rearing is one of the most sensitive activity. Larval rearing in each tanks is separately monitored. The nauplii from the spawning tank collected in the harvesting bucket is estimated for population using sampling method before stocking. Mean value of atleast three samples of 50ml from the bucket is considered for the estimation, using the formula below:

$$\text{Total nauplii in the bucket} = \frac{\text{Mean no. of nauplii in the sample}}{\text{Volume of the sample}} \times \text{Volume of water in the bucket}$$

The nauplii are stocked in larval rearing tanks normally 60 to 100 numbers per litre. The larvae are reared upto PL4/PL5 for about 15 days clean healthy environment is needed for the rearing which is maintained through water exchange daily for zoea stage. The feed are provided *Artemia* nauplii and micro algae. The survival rate upto PL5 stage is around 50 per cent. When the attack of any disease occur, antibiotics and other medicines of aquatic grade are used.

5.7 Post larval rearing

Post larval rearing is done upto PL20 stage which takes about 20 days PL20 is then harvested for selling. Post larvae of good quality, active and healthy are stocked in the indoor PL rearing tanks 25–50 PL/Litre and 8–10 in outdoor PL rearing tanks. Algal feeding is continued @50,000/ml concentration. *Artemia* nauplii is added 2 nauplii PL. In addition egg austered is given up to PL5. From PL6 and above pelted feed is given @ 2–3 per cent of the biomass. Water exchange is required upto 80–100 per cent per days. If any disease problem is seen, the post larvae are feed with antibiotic with egg lusted or aquatic grade antibiotic/ medicines are directly applied.

5.8 Counting and Packing

Seed is counted and packed in polythene bag with 1/3rd water and 2/3rd oxygen. Depending on the distance for transportation of seed (PL20) packed with a packing density 2000–5000 PL in 35 lit capacity bag. Ice filled packets are kept in the four corners of the thermocol packing inside paper cartons which are tightly sealed.

5.9 Larval or Post Larval Feed Preparation: *Artemia* Cyst Hatching Techniques

Artemia sp. is found or raised in the brine ponds. It produces encysted embryo at gastrula stage when hydrated and suitable temperature and aeration provided, hatch out into nauplii. The daily requirement of hatch out nauplii from the cyst is needed which is fed to the larval tanks in the hatchery. *Artemia* cyst is marked by several companies. Continuous aeration and 40w tube light fixed at 20cm above the tank are provided. The chlorine treated (50 ppm a few second) Artemia cyst is released into the Artemia hatching tank. Hatching is done 1–2 gm cyst per litre upon hatching (24–36 hours after introduction) the aeration is stopped and a focus of light beam is set near the glass window of the tank. The nauplii are then concentrated at the light which are siphoned out to the harvesting bucket. The nauplii are washed thoroughly and used as feed to the larval or post larval tanks as per requirements.

5.10 Larval or Post–Larval Feed Preparation: Micro Algal Culture

Now the micro algae being cultured for the feeding of shrimp crustacea larvae (shrimp larvae) which is related to the success of any hatchery operation (Gopinathan, 1995).

The culturable micro algae are diatom like *Chaetoceros, Cyclotella, Skeletonema, Costinodiscus* and some flagellates like *Isochrysis* and *tetraselmis, Monochrysis, Dicrateria.*

5.10.1 Preparation and Selection of Culture Media

A number of media are now a days for culturing the different groups of micro algae which are the following:

a) Schreiber's Medium

This solution has the following composition.

Filtered sea water 1 lit.

Sodium nitrate – 0.1 gm.

Sodium acid Phosphate – 0.02 gm.

Soil extract – 50 cc.

The soil extract is prepared by dissolving 1kg of good fertile soil (garden soil) in 1lit. of distilled water and autoclaving it for about 30 minutes under the pressure of 15lb/m^2. After one day, the supernatant soil extract is decanted into a flask and auto calved again for 30 minutes. The freshly prepared soil medium is always preferred. This medium is used for culturing almost all microalgae.

b) Miquel's Medium

Solution A

Potassium nitrate – 20.2 gm

Distilled water – 100 ml.

Solution B

Sodium or thiophosphate 4 gm

Calcium chloride – 2 gm

Ferric chloride – 2 gm

HCI – 2 ml.

Distilled water – 100 ml.

This medium is prepared by adding of 0.55ml of solution A and 0.5ml of solution B to one litre of filtered and sterilized seawater. This medium can be used for culturing of various type of micro algae.

c) 'Conway' or Walne's Medium

This medium is largely used for culturing phytoflagellates which has three stock solution viz. A, B and C whose composition is given below:

Solution A

Sodium nitrate or potassium nitrate 10 gm

Sodium phosphate 2 gm

Ferric chloride 0.13 gm

Manganese chloride 0.04 gm

Boric acid 3.3 gm

EDTA 4.5 gm

Dissolve in 100 ml of distilled water

Solution B

Zinc chloride 2.1 gm

Calcium Chloride 2 gm

Ammonium molybdate 2 gm

Copper sulphate 2 gm

Dissolve in 100 ml of distilled water

Solution C

Cobalamine (vitamin B_{12})

Thiamin (vitamin B_1)

Dissolve in 100ml of distilled water

This medium is prepared by adding 1 ml each of the above solution to 1 lit. filtered seawater.

d) TMRL (Tung Kang Marine Research Lab) Medium

This medium is chiefly use for mass culture of diatom (*Chaetoceros, coscinodiseas* etc.) is prepared as follows:

Potassium nitrate 1 gm

Sodium phosphate 0.1 gm

Ferric chloride 0.03 gm

Sodium silicate 0.02 gm

The medium is prepared by dissolving all those things in 10 lit. of filter seawater.

e) Mass Culture Medium

For mass culture of diatoms in open circular or concrete system, the following medium is generally preferred.

Ammonium sulphate 2 gm

Urea 150 mg

Ferric chloride 200 mg

Sodium silicate 200 mg

Potassium phosphate 200 mg

Now these are dissolved in 100 lit. of filtered sea water.

5.10.2 Method of Isolation

The isolation of the required species for inoculation in the culture medium can be done by the following methods:

a) Pipette Method

Larger organisms can be pipetted out using a micro–pipette under a microscope and washed them. Then transferred to culture petridishes or tubes having suitable sterilized culture medium (Fox, 1983).

b) Centrifugation Method

By centrifuged at 3000 rpm sediment organisms are washed and transferred into a sterilized petridishes of tubes containing the autoclaved medium.

5.10.3 Culture Method

A) Small Scale Culture Method

a) *Agar Plating Method:* For preparing the agar medium 1.5 per cent agar is added to 1 lit of desired medium or filtered seawater and this solution is autoclaved for about 15 minutes under a pressure of 15 LB/ cm^2. This agar medium is then poured aseptically into the sterilised petridishes and left undistributed for 24 hours. Instead of petridishes, test tubes may also be used.

   Now the isolated species are picked up by a platinum loop or needles under the microscope and streaked on the surface of the agar plates. After inculation, these petridishes are placed in an incubation chamber maintaining constant temperature (25°C) and light intensity (1000 lux) for 7 – 15 days. When the individual cells multiply and leading to formation of colonies. Now these colonies could be removed by platinum loop and transferred into culture tubes (Santaram *et al.* 1987).

b) *Direct Culture Method:* The isolated and washed microalgae cells may be directly inoculated to the sterilized test tubes containing the medium.

B) Mass Scale Culture Method

After culture of small scales, the tubes containing the culture may be transferred to sterilized culture flask 1 (lit). Then from flasks the medium will be transferred into carbuoys (20 lit) containing medium. For out door culture, 200 litre fiber glass tanks are used and lastly for mass production, 1 ton fiber glass tanks or 1–6 tons cemented tanks will be used.

5.10.4 Factors Affecting the Micro Algal Culture

a) Light : For maintaining the culture 500 lux light is essential. While for mass culture 2000–3000 lux is necessary. A photoperiod of 12 h light: 12 h dark is ideal for maintaining the stock culture

b) Temperature : Normal room temperature (26–30°C) is not ideal for maintenance and culture of micro algae. So A.C. rooms are used for keeping the stock culture as well as mass culture. In there temperature should be 23–25°C during daytime in night A.C. can be switched off for few hours.

c) Aeration : It was noticed that if aeration has given mass culture tanks, culture will remain in the growing phase 3–4 days more than tanks with aeration. Aeration not only helps culture for oxygen but also supply CO_2 required for photosynthesis lastly aeration will prevent the setting out the cells at the bottom of the culture tanks.

d) Anticontamination : All working with the micro algae should be done with one species at a time culturist should wash their hands after working with one species of algae. All regular pipettes glass tubing and glass wares should be sterilized before use.

5.10.5 Growth Phase of Micro Algae

In laboratory culture, under controlled conditions, increase in cell numbers follows a characteristic pattern in which, the following phases are noted

a) Lag Phase or Induction Phase : This phase is characterised by the absence of any increased cell number as there will be no cell division for few hours.

b) Log Phase or Exponential Phase : In this phase the cells start multiplication and grow rapidly.

c) Stationary Phase : Here the cells are stationary without further any cell division.

d) Death Phase : In this phase, the cells start decaying and the cells may lose its viability and start to die.

5.10.6 Harvest of Culture

The fully grown culture should be harvested during the exponential phase of the micro algae after determining the concentration. If the culture has entered into the stationary or declining phase, the metabolites will be very high and cells may not be in healthy condition. The rearing larvae may not show the expected growth if fed with this feed. It is also better to harvest the cells of tube or flask during exponential phase and transfer them to carbuoys or out door system for mass production.

5.10.7 Counting of Micro Algae

After sampling, sample has to be treated with eosin or Formalin to kill the cells. Now one drop is taken through sterilised pipette and counting is done by haemocytometer.

$$\text{Algal Density} = \frac{\text{Average No. of Cells in Upper Block} + \text{Average No. of Cells in Lower Block}}{2} \times 10^4$$

5.10.8 Stock Culture Maintenance

Stock culture of micro algae are maintained in a special air conditioned room nearer to mass culture. The autoclaved seawater is poured into the Haufkin culture flasks with algae and required nutrient medium will be added. When the maximum exponential phase is reached, the tube light can be reduced for further growth. In the stationary phase when the colour of the culture turns brown the micro algae can be kept for a period of 2 months in the stock culture room, under controlled conditions of light and temperature, with or without aeration.

5.10.9 Preservation of Micro Algae

Preservation of culture media is needed for adverse weather conditions or shutdown period of the hatchery. Preservation is done by freezing or sundrying.

At first the arising of the culture medium is done and the culture is to be left for one hour. After one hour the algal mass is deposited at the bottom of the tank. Now the algal mass is harvested which is ready for freezing or sundrying.

a) Sundrying : Drying of the algae can be done by pouring the mass in white enamel trays and keeping the same in high sunlight. When the algae has dried up thoroughly scraping of the powder from enamel tray is kept in glass bottle.

b) Freezing the Algae : Before freezing the algal mass, some preservative like dimethylsulphoxide or glycerol (a few drops) can be added and then it is poured into polythene bags. The bags are kept in deep freezers.

5.11 Post Larval Feed: Dry Pelletised Feed

This feed can be prepared in the hatchery using a mix of dry ingredients like shrimp meal, fish meal, mussle meat, liver oil, minerals, vitamins, cereals etc. These pelletised meal is used for post larvae instead of *Artemia* nauplii which is habituated them for prepared feed like grow out pond later.

Table 1: Diseases in Shrimp Hatchery and Thier Prevention of Control

Diseases	*Pathogen*	**Symptoms of Larval Stages and Their Affected Parts**	*Prevention or Control*
Viral Disease			
i) Baculovirus midgut gland necrosis (BMN)	Baculovirus	Infection in hepatopancreas and midgut in all stages	There is no chemical or antibiotic to control into use of diseases free stock; discard fully infected stock
ii) Baculovirus Penaei (BP)	Baculovirus	Infection in hepatopancreas and midgut in all stages	There is no chemical or antibiotic to control this disease prevention method is to use of disease free stock, discard fully infected stock.
iii) Infection of hypodernal and haemopoietic necrosis (IHHN)	Picorna like virus (IHHN virus)	Infection of hypodermis and haemopoietic organ.	Use of disease free stock, discard fully infected stock.
Bacterial Disease:			
i) Bacterial necrosis	Miscellaneous bacteria	Discolouration of appendages or localised necrosis is found in zoea, mysis and PL stage	Control method is using of prefuran 1.0 ppm or malachite greed 2.0–4.0 ppm. Prevention method is to avoid over crowding, feed well, and maintain good water quality.
ii) Vibrosis	Vibrio sp.viz. Vibrio parahaemolyticus v.alginolyticus, v. angvillarium	Development of yellow colour, later turn to white. liver, haemolymph and midgut is turbid which is found in mysis and PL stage.	Control method is to use of prefuran 1.0 ppm or flurazolidone 8.5 ppm. Prevention method is to avoid overcrowding, feed well and maintain good water quality.

Contd......

Diseases	*Pathogen*	**Symptoms of Larval Stages and Their Affected Parts**	***Prevention or Control***
iii) Luminescent vibrosis	vibro harveryi v. splendious v. fischeri	Development of redish colour, show blue green luminescence is total darkness. Haemolymph, gut and haepatopancreas get affected in mysis/PL stageSeen back to gill filaments, chocked gills in zoea/mysis/PL stage.	Controlling method is use of chloramphenicol 10.0 ppm, or erythromycin 2.0-4.0 ppm, or prefuran 1.0 ppm, furazolidone 3.0 ppm. Prevent over crowding and exchange water.
iv) Filamentous bacteria	Leucothrix mucor	Seen back to gill filaments, chocked gills in zoea/mysis/PL stage.	Use of prefuran 1.0 ppm, or methylene blue 8.0-10.0 ppm, prevent over crowding and exchange water regularly.
vi) Shell diseases	Chilinoverous bacteria	Tips and edges of exoskeleton get eroded and subsequently enter the body of the PL.	Use formalin 25.0 ppm, prevent over crowding and exchange of water.
Fungal Disease			
Larval mycosis	*Lagenidum Sp.*	Entire body space (inside and outside) gets filled with fungal hyphae in Nauplii/Zoea/ Mysis.	use traflon 0.1 ppm or malachite green 0.0075 ppm. Prevent over crowding.
Parasitic Diseases			
i) Protozoan infection	*Vorticella* sp. *Zoothamnium* sp. *Epistylis* sp. *Acinata* sp.	Choke the gill, affect the locomotion, moulting and feeding	use of cupric chloride 0.25-1 ppm, or Formalin 15.0-20 ppm or cutine plus 0.25-1.0 ppm or malachite green 0.0075 ppm. Exchange water daily.

5.12 Diseases and Their Control

Diseases attack is a natural phenomenon in the hatchery system. The medium of rearing tank encourages the pathogens to proliferate when the larvae grown under dense population with high concentration of feed. The infection is by the precautionary measures like maintaining the water quality, general hygiene and a systemic prophylactic disinfection schedule using antifungal, antibacterial drugs. Maintaining proper chlorination, filtration through sand filter and cartridge filters can improve water quality. Treatment with EDTA reduces turbidity and heavy metals. Treatment of zeolite reduces the toxic gases. Maintaining the temperature 28^0–32^0C also avoid most of the bacterial infection which is common in cold months. Precautionary measures like water disaffection, avoiding use of aged algal culture, inadequately disinfected *Artemia* nauplii can reduce or avoid disease problems. Prevention of diseases is necessary since the treatment has very little effect on the pathogen as the pathogen strains tend to become immune to prescribe medicine or chemicals which are use for a long time in a particular site. The general disease, symptoms and their prevention measures or control is given in table 1.

Chapter 3

Culture Technology of Tiger Prawn, *Penaeus monodon*

1. Introduction

Marine shrimp (*Penaeus monodon*) farming is a century–old practice in many Asian countries. Until a decade ago, this commodity was generally considered a secondary crop in traditional fish farming practices. Shrimp fry trapped in coastal paddy fields or brackish water fishponds are allowed to grow upto marketable size and harvested as secondary crop. However, in recent years when higher income is derived from harvest of shrimp than the principal crop, many farmers have converted their rice fields, salt beds and fishpond into shrimp farms.

In the traditional farming system the ponds are stocked with fry either collected from the wild or concentrated through tidal water entering the ponds. Shrimp production is inconsistent and varies from year to year due to the dependence on seasonal supply of fry from the world.

In India scientific shrimp farming started during 1980 after hatchery seed production technology evolved in Indian conditions by the central Marine Fisheries Research Institute at Cochin, demonstrated the possibilities for production of good quality shrimp seed. Massive tiger shrimp seed production stated by Marine Products Exports Development Authority (MPEDA) through their hatchery with 40 million annual seed production capacity in Andhra at the village Mangamaripeta near Vishakhapatnam. This development encouraged commercial seed production and farming activity. The success in mass production of hatchery–bred tiger shrimp fry has accelerated shrimp farming development in Andhra, Tamil Nadu, Orissa and West Bengal.

Tiger shrimp farming has now developed into an important export–oriented food industry especially in South Asian countries. The perception of an unlimited market demand, high export price, generation of employment and increase in foreign exchange earnings may have encouraged many countries in the region rich in aquatic resources to place high emphasis on the development of the shrimp culture industry.

2. Site Selection

The selection of a suitable site always plays a major role in shrimp farming. The determination of a site of *P. monodon* farming is made only after through analysis of information topography, ecosystem, meteorological and socioeconomic conditions in relation to farm design and economic viability. Criteria are given below:

2.1 Water Quality

Water quality includes all the inherent physiochemical and microbiological characteristics of water. pH is generally considered as one of the most important factors. In any chosen site, the pH of the water preferable range from 7.5 to 8.5.

The other quality important characteristic of water is the level of oxygen saturation throughout the water column. Fluctuation in dissolve oxygen should be predetermined and oxygen level is preferably not lower than 4 ppm.

Salinity variation is one of the important factor for tiger shrimp production. The tiger shrimp grows faster at 15–30 ppt. So, salinity should remain uniform at normal weather and should not drop abruptly during rainy days.

The water not be too turbid. The water is preferable to be rich in microorganisms.

2.2 Tidal Fluctuation

The tidal characteristics of the proposed site should be known knowledge, of this parameter is of extreme importance in determining pond bottom elevation of dike, slope ration and drainage system.

Areas best suited from shrimp farming should have moderate tidal fluctuations preferable 2–3 meters. In areas where the tidal range in greater than 4 meters, the site may prove uneconomical to develop or operate a farm, because high pond dikes will be required. In those areas where tidal range is less than one meter, water management will be expensive requiring the use of pumps.

2.3 Soil

The type and texture of the soil of the area should be analysed before setting on a site for shrimp farming. Soil samples must be taken at random location (preferably upto a depth 0.5 meter) and subjected to physical and chemical tests to determine the acidity, amount of organic load, level of fertility and physical composition.

The soil at the proposed site should have enough clay content. This is to ensure that the ponds constructed will hold water. Good quality dykes are usually built from sandy clay or sandy loam materials, which harden and easily compacted.

2.4 Topography

It is essential to have detail topography of the selected site for pond design and far layout. coastal sites where the slopes run gently towards the sea are easier for pond development requiring less financial inputs since escavation is minimal. Filling and draining of water like wise is easily facilitated by gravity.

In the areas where the above conditions are not available, the use of mechanical pumps may be resorted. In this case topography related constrains are the availability of sufficient quantity of soil for dyke construction that may be obtained from excavation of ponds. It may prove uneconomical if the dyking material to be transported from the outside the chosen area.

2.5 Source of Seed

Close proximity of the site to the fry ground (wild or hatchery) is advantageous in that the post larvae being collected for stocking are not subjected to too much transport and handling stress.

2.6 Accessibility

Accessibility is an important consideration in site selection. Overhead cost and delay in the transport of materials and products can be minimized.

2.7 Other Factors

Other consideration should be given to a number of farm related factors which are given below:

a) Good electricity connections and supply.

b) Good availability of labour.

c) Reasonable road and transportation to the site.

d) Reasonable communication.

e) Availability of formulated and natural feed.

f) Availability of equipment supply and maintenance.

g) Proximate processing facilities.

Site selection is an intensive process and is time consuming. However, for semi intensive farming one cannot afford not to have control over all parameters that affect growth, animal health and feed efficiency. Many of the larvae in certain sites can be overcome by proper engineering.

3. Design and Construction of *P. monodon* Culture Pond

There is no standard design for a shrimp rearing pond. The improved structural design is largely due to the behaviour of the penaeid shrimps. *P. monodon* is a benthic animal and it has a habit of gathering along the tank walls. So, any design to increase the wall surface e.g. adding substrates or an elevated earthen platform extending along the edges of a pond, can promote high stocking densities.

An ideal shrimp farm is a complex, which consisting of following parts:

a) Various size of pond for nursery and growout.

b) Water control structure including embankments, supply and drainage canals and sluice gates.

c) Other facilities such as roads, living quarters, workshops and warehouses etc.

3.1 Layout of Culture Ponds

Rectangular or square ponds are appropriate for tiger shrimp culture, but the rectangular pond is more preferable. The longest axis of a pond should be parallel to the prevailing wind direction, which facilitates water movement generated by wind action. So dissolve oxygen will be increased that minimized the water temperature fluctuations in summer months. Due to different types of culture, the following various size of ponds are recommended:

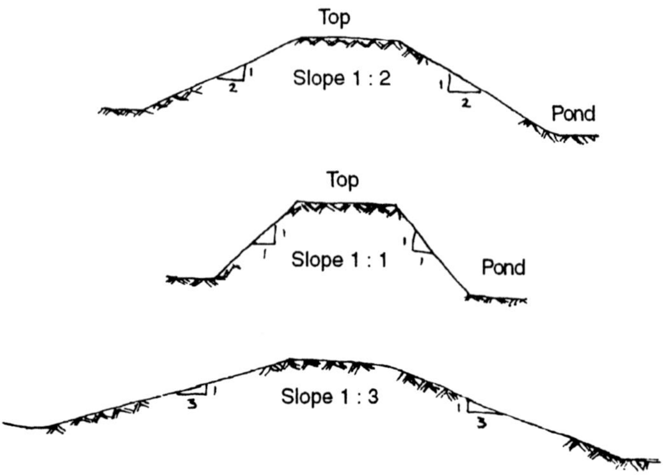

Fig: Typical Slope of Dike.

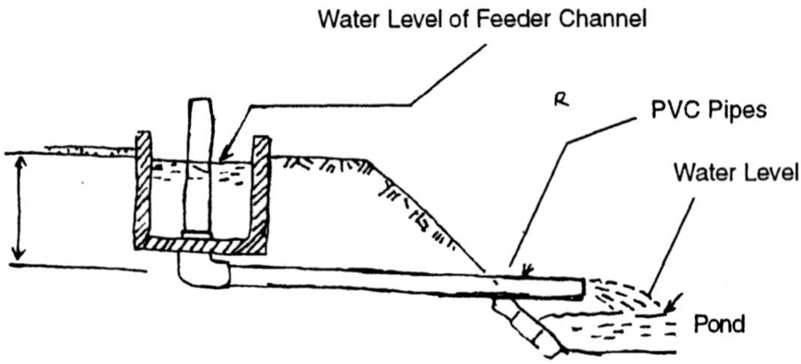

Fig: PVC Pipe Inlet.

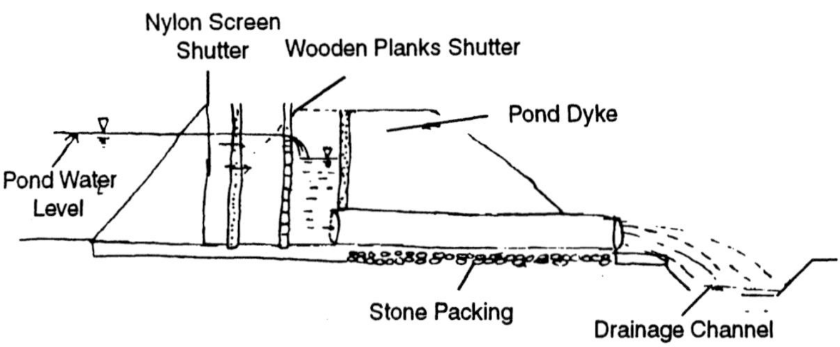

Fig: Pond Outlet.

Nursery pond	500 to 1,000 m^2
Growout pond	
intensive	0.25 to 1.0 ha
semi intensive	0.5 to 2.0 ha
extensive	1.0 to 10 ha

The average depth of growout pond is 1.0 to 2.0 meters. Pond bottom should be free from projecting rocks and tree stumps. The bottom must have a gradual slope from inlet gate towards the drainage gate. The suggested ratio of the slope is 1:500.

3.2 Dykes

Dykes not only serve as boundaries to indicate pond size and shape but also function to hold water within the pond as well as protecting other farm facilities from flood. Dyking materials must be tested for load bearing capabilities. Design and construction of embankment must be based on sound engineering principals and economic feasibility.

To computer for the height of the dyke, the following formula could be used:

$$H = \frac{(H_W - G) + FB}{1 - \% \text{ of Shrinkage}}$$

Where,

H = height of designed dyke

H_W = highest height water level from past record

G = ground level over mean sea level

FB = height of free board

The slope of dyke is maintained at an average ration of 2:1 to 3:1. Slope of a dike also highly depends on soil equality. For good clay soil, the recommended slopes are:

1:2 when dyke height is above 4.26 m and exposed to wave action.

1:1 when dyke is less than 4.26 m and the tidal range is greater than 2 meters.

2:1 when the tidal range is 1.0 m or less and the dyke height are less than 1.0 meters.

The crown of the dyke between ponds should be 1–5 meters.

The soil erosin of the dykes may be controlled, if the fast growing grass species are planted on the dykes.

3.3 Water Control Gate (Sluice Type)

Water control gates are often called sluice gate. When designing a sluice gate, it is essential to consider tidal fluctuations and gravity in order to ensure effective control of the inflow and outflow of water within a give period.

Irrespective of the material to be used to construct the water gates (e.g. wood, ferrocement, and reinforced concrete), the following requirement should be met:

a) A gate should have adequate capacity for the required amount of water to be taken in or drained out.

b) A gate should be constructed in a position that water can be totally discharged.

c) A gate should have sufficient grooves for placement or filter screens, slabs and harvest nets.

d) A gate should be firmly places at the pond bottom and properly linked with the dikes to prevent seepage and possible collapse.

A standard gate design consists of tied stern wing, side walls and bed structure. Grooves for slab are usually set at the inner side of the gate..

The size of the gate is based on the total water requirement of a pond. Water intake volume is calculated using the equation:

$$Q = CA\ [2g(H\text{–}h)]^{1/2}$$

where,

Q = rate of flow (m/sec).

C = cross section of the flux (calculated by multiplying the width of gate opening and its depth).

A = coefficient of discharge (0.61).

g = gravitational constant (9.8m/sec^2).

H = tidal level of the river or sea.

W = water level in the canal or pond.

3.4 Supply and Drainage Canal

It is necessary to construct supply and drainage canals for

supply and drainage of water for shrimp pond. So, a pond must be possess separate canals for drainage and supply of water. Both supply and drainage canals would likewise serve as water level control in the pond. It is important that the siting of the canal systems takes advantage of the natural waterways within the proposed site.

Dimensions of supply and drainage canals are calculated by using the following equation:

$$Q = AV$$

where,

Q = volume of water discharge.

A = cross–section area of the canal.

V = velocity of water flow.

V can be calculated by the following formula:

$$V = R^{2/3} \times S^{1/2} \times 1/n$$

where,

R = depth of water flow.

S = canal bed gradient.

n = coefficient of roughness (0.02).

4. Culture Operation and Management

Culture operation of tiger shrimp in artificial ponds can be divided into following steps :

4.1 Pond Preparation

In any earthen pond culture system, the bottom soil plays a major role in pond yield. High organic matter content in neutral soil often promotes higher primary productivity and hence higher fish yield. So, pond preparation is a first step towards ensuring a better pond production.

4.1.1 Soil Sampling

Prior to pond preparation, soil samples are collected from the pond bottom for pH and organic matter contents analysis. Soil pH analysis is generally conducted to determine lime requirement.

In most soil laboratories, sample analysis takes one to two weeks before the result can be given. To prepare for the analysis, the air dried soil sample must be pulverized using a wooden mallet of mortar and pestle. It is then passed through a 2 mm mesh sieve and stores in a property labeled bag or glass jar. Now, the sample is ready for pH determination.

4.1.2 *Leaching*

When the pond soil is found to be acidic, it is normally leached. This is done by flushing and washing the pond bottom with water to leach away undesirable metallic compounds like aluminium, iron and excess sulphur ions.

4.1.3 *Drying*

After flushing out the pond, all inlet and outlet structures are sealed and the pond is allowed to dry, a minimum of 7–10 days, or until the pond bottom surface can support the weight of a person without setting and the soul surface has opened in deep cracks. Drying the pond facilitates:

a) Aerating of surface sediments, oxidizing reduces compounds such as hydrogen sulphide, nitrite, ammonia, ferrous ion, methane etc., which are toxic to shrimp.

b) Decomposition and mineralization of organic matter.

c) Disinfection of the pond bottom.

d) Ploughing/tilling of the pond bottom to expose underlying sediment layers that may be anaerobic and reduced.

e) Elimination of undesirable mats of filamentous algae.

f) Elimination of fish eggs, crab larvae and potential predators.

Drying is a very important step in the pond conditioning process, however, excessive drying is counter productive and reduce pond turnover and total corps per year. If the stocking programme dictates that a pond must remain empty for more that three weeks (excluding preparation time), it might be advisable to refill it for a few days to wet the bottom and then dry again.

4.1.4 *Tilling*

Tilling or ploughing of bottom soil improves soil quality by exposing subsoil to a depth of 5–15cm suing a plow, thereby speeding up the oxidation process and the release of nutrients that are locked in the soil. Double tiling ensure through disintegration of soil agglomerates.

4.1.5 Liming

Liming is the application of calcium and magnesium compounds to the soil for the purpose of reducing soil acidity. It is usually applied after the pond drying stage.

The favorable actions of liming are:

a) Kills most microorganism especially parasites due to its caustic reactions.

b) Raised pH of acidic water to neutral or slightly alkaline value.

c) Increase the alkaline reserve in water and mud, which prevents extreme change in pH.

d) Neutralizes the harmful action of certain substances like sulphides and acids.

e) Promotes biological productivity since it enhances the breakdown or organic substances by bacteria creating a more favorable oxygen and carbon reserves.

f) Precipitates suspended or soluble organic materials, decreases biological oxygen demand (BOD), increase light penetration, enhances nitrification due to the requirement of calcium by nitrifying organism.

h) Indirectly improves fine texture bottom soil in the presence of organic matter.

However, excess liming can be damaging because it decreases phosphorus availability through precipitation of insoluble calcium or magnesium phosphate.

Agriculture limestone ($CaCo_3$), calcium oxide (Cao), calcium hydroxide [$Ca(OH)_2$] and mixed calcium magnesium carbonate [$CaMg(Co_3)_2$] are generally applied for liming.

Following inspection of the pond bottom, agriculture limestone ($CaCo_3$) should be applied if soil pH reading is less than 7.0.

When it is desired to sterile pond bottoms, burnt or hydrated lime [CaO or $Ca(OH)_2$], may be applied at 1–2 tons/ha. This is thought to raise the pH of the soil enough to kill pathogenic bacteria, which can reside in soils between crops.

Some farmers apply hydrated lime [$Ca(OH)_2$] or quicklime (CaO) because they react faster, possess approximately 35 per cent and 75 per cent respectively more neutralising activity than $CaCO_3$ and also act as disinfectants and pesticides. It is generally not recommended to use Ca(OH)2 or CaO for soil pH neutralization. But exceptions to this rule are ponds with acid sulphate soil or ponds in which disease out breaks have led to poor harvest results of previous production cycles.

Mixed calcium, magnesium carbonate $CaMg(CO_3)_2$ is also known as agricultural limestone. The carbonates are the least reactive.

Liming can be done in three different ways:

a) by broadcast over dried pond which includes the dyke walls.

b) by mixing with water and spraying over the pond.

c) by liming the water flowing into the pond.

Agricultural limestone of the highest available purity should be utilized. Tiger shrimp farmers should rely on following doses:

pH	Lime (Mt/ha)
4	5.0
4.5	4.5
5.0	3.75
5.5	3.25
6.0	2.50
6.5	1.90
7.0	1.25

4.1.6 Water Filing

Fill pond 30 cm with 25 ppt salinity and drain after 3 days. The refilling of water upto 30 cm is practice, recommended specially for semi intensive culture system. The cycle of dry–plow–fill–drain treatment will improve oxidation state of the pond bottom.

4.1.7 Application of Inorganic Fertilizers

It is recommended to adopt the practice of incorporation urea of some other nitrogen source into the soil in order to accelerate organic matter decomposition. A dose of 50–100kg urea/ha may be used before refilling the pond to 50cm.

4.1.8 Refilling the Water

Add 10cm of water everyday until the pond water is 100–120 cm deep. Urea and NPK fertilizer may be added as required till the colour of water is brown with a yellowish hue.

4.2 Selection of Shrimp Fry

The most important criteria for selection are the stage of development of the fry. Identification of anatomical parts of the fry will require a microscope but a farmer could do well to carry a small eye piece/magnifying lens to the hatchery for observation of the fry before purchase. It is recommended that farmers should prefer to take PL20 animals.

4.3 Transportation of Shrimp Fry

If wild fry collection centers of hatcheries are located far away from ponds and fry have to be transported to pond area, current industry practice is to pack 2,000–2,500 post larvae in plastic bags which are oxygenated, sealed and packed in cardboard cartons lined with thermocol. It transportation time to the ponds is more than six hours, it is recommended that the temperature is lowered to between 20–26^{0}C by adding small bags with ice on top and around the bags containing the fry. Reduction in temperature should also be gradual to avoid any stress on the animals. Temperature reduction is done to reduce the metabolic rate and oxygen consumption.

4.4 Acclimatization and Stocking of Fry

The best stocking time is early in the morning or after 6 pm at night. The collection and transportation of seed from the hatchery, coordinated to match the stocking time. The stocking density of fry *P. monodon* in different culture operation is given below:

Culture Practice	Stocking Density (per m^3)
Extensive	1–5
Modified Extensive	6–15
Semi Intensive	16–50
Intensive	51–75
Super Intensive	76–300

It is advisable to postpone delivery of fry from the hatchery if the alga bloom in the pond is not correct. The seeding process i.e. acclimatization of seed is follows:

a) Empty the bag in a plastic container on basin.
b) Observe the fry for level or activity and mortality.
c) Lower the basin till it floats in the water. The water of the container should not mix with the pond water.
d) Hold the basin with one hand, and add water of the pond into the container slowly. After sometime tilt the basin as the fry should slowly swim out into the pond water.

If fry don't swim out quickly then differences in pond water parameters and the hatchery are more require acclimatization.

4.5 Culture Technique

Tiger shrimp culture operation in Asia can be grouped into three major categories:

a) *Extensive culture:* Those that depend entirely on natural food (Stocking density 1–5m^2).
b) *Semi Intensive:* These that depend on both natural food and supplementary feeds (Stocking density 16–50/m^2). Upto 15 per cent water may be exchanged when required.
c) *Intensive Culture:* Those that entirely depend on artificial diet (Stocking density 51–75/m^2). 15–50 per cent water may be exchanged when required.

The following figure will clear the culture technique:

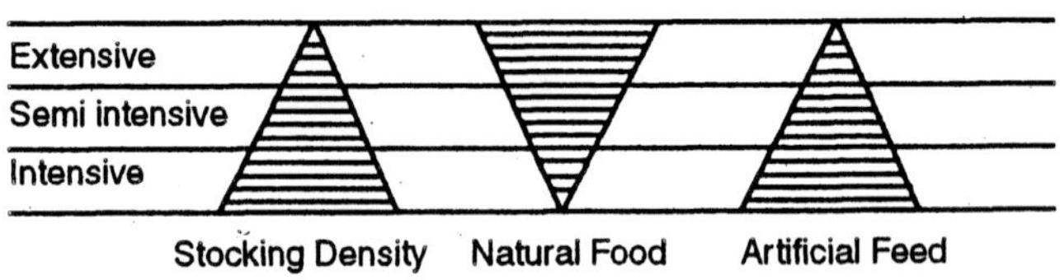

Other than these culture operations, sometime modified extensive and super intensive culture technique is to be used. Modified extensive culture technique depends on mainly natural food but sometime artificial feed may be provided. Stocking density is to be maintained 6–15/m^2.

But super intensive culture entirely depends on artificial diet. In there, 50–100 per cent water may be exchanged per day. Stocking density is to be maintained 76–300/m^2.

4.6 Water Quality Management

In any shrimp farming, management of water quality is of primary consideration. Particularly in ponds with higher stocking rates. Degradation of water quality is detrimental to shrimp growth and survival. Good quality water is usually defined as the fitness or survivability to the water for survival and growth of shrimp.

4.6.1 Salinity

Younger shrimps appear to tolerate a wider fluctuation of salinity than the adults do. The post larvae of many penaeid species can tolerate wide salinity fluctuation, which has little effect on their survival or growth. In pond condition, *P. monodon* can tolerate wide range of salinity from as low as 5 ppt to a high 40 ppt.

Due to evaporation rate in some countries, salt concentration in ponds gradually increases during the summer months. In such cases, the water should be changed frequently either by pump or through tidal exchange.

4.6.2 Temperature

Water temperature plays a very important role in regulating the activities of culture animal. In case of *P. monodon* the optimum temperature is about is 25–30^{0}C. Higher temperature may be maintained through water exchange.

4.6.3 Dissolved Oxygen (DO)

Maintenance of adequate level of dissolved oxygen in pond water is very important to shrimp growth and survival. Prolonged exposure to the stress of low concentration of oxygen lowers their resistance to disease and inhabits their growth, often resulting in mass mortality. This is particularly common in intensive operation.

Dissolved oxygen in the pond water comes from two sources – by product of photosynthesis and from the diffusion of atmospheric air. At night, both plants and animals continue to respire while oxygen is being added to the water from the atmosphere, cause total depletion of DO (specially at dawn).

Depletion of DO in pond can be controlled by the following measures:

a) Water exchange through renewal of pond water with fresh water either by tidal flow or pumping.

b) Installation of aeration system. But it is essential to consider maximal utilization of natural environment (wind action & water movement, maximum contact of water surface with atmospheric air etc.) to maintain higher dissolved oxygen content in pond water.

Aeration: Paddle wheels are commonly used in semi intensive prawn culture and is one of the major capital cost item in the farm. The paddle wheel aerators are used to increase contact surface of water with air there by increasing the area through which oxygen is absorbed by water to create a circular movement of the pond water.

There is no hard and fast rule regarding the number of paddle wheels to be used in a pond because it is difficult to define the relation between biomass in the pond and the interaction of various water parameters. However, according to experience, each HP paddle wheel can yield about 1,000 kg of marketable size shrimps. Use of less than four aerators is not recommended.

The arrangement of paddle wheels in the pond is done to maximize circular efficiency and minimize dead corner area. Aerators are usually arranged parallel along the banks and 5 to 10m distance from the dykes depending upon the pond size. The pond water movement through aeration has the following advantages:

a) Increases the dissolved oxygen level of the water and prevents oxygen depletion during the night.

b) It accelerates the diffusion effect of not only the oxygen, but also enables the release of CO^2.

c) It facilitates the volatilization of undesirable gases such as N_2, NH_3, H_2S etc.

d) It reduces the daily fluctuation range of pH value.

e) It helps in mixing the pond water and maintenance of ideal conditions of water of the correct column.

f) As the pond water moves in a circular fashion the pond bottom is cleaned, and the waster matter gets accumulated in the centre and the corners. By this method most of the pond bottom is kept clean.

4.6.4 pH

The pH of the pond water is indicative of its fertility or potential productivity. Water with pH ranging from 7.5 to 9.0 is generally regarded suitable for shrimp production. The growth of shrimps is retarded if pH falls below 5.0 which can be corrected by adding lime to neutralized the acidity. Water of excessive alkalinity (pH value 9.5) may also be harmful. In ponds, which are excessively rich in phytoplanton, the pH of pond water usually exceeds 9.5 during late afternoon that can be corrected by water exchange.

4.6.5 Nitrogen Compound

Nitrogen in pond exists is different forms such as nitrate, nitrite, ammonia and various forms of organic nitrogen. In pond culture activities, ammonia nitrogen (in the form of un–ionized ammonia) is considered important since this compound is toxic to aquatic animals at certain concentration. The pH and temperature of the water regulates the proportion of total ammonia, which occus in un–ionized form. Studies have shown that exposure to ammonia concentration of 0.45 mg NH_3–N/lit would reduce shrimp growth by 50 per cent.

This compounds can be corrected by water exchange.

4.6.6 Hydrogen Sulphide (H_2S)

Hydrogen sulphide can severely affect shrimp growth, which is produced by chemical reduction of organic matter that accumulates and forms a thick layer of organic deposits at the bottom. At levels of 0.1–0.2 ppm in the water, the shrimps appear to lose their equilibrium and die instantly at a concentration of 4 ppm.

Using iron oxide (70 per cent ferrous oxide) to treat the bottom soil containing high level of H_2S is not economical. The cheaper means is by frequent exchange of water to prevent building up of H_2S in the pond.

So far, water exchanges is still the most effective and widely employed method to maintain good water quality. In normal course, recommended daily water exchange schedules are given below:

Culture Period (in days)	Salinity (in ppt)	Water Exchange Rate (per cent in daily)	Water Depth (cm)
0–15	25–30	2.5	80
16–30	20–25	5–8	90
31–45	20–25	10–12	100
46–60	15–22	12.5–15	120
61–90	15–20	15–20	120
91–122/till harvest	20–25	20–25	150

4.7 Feed and Feeding Management

One of the most important operational functions in tiger shrimp culture is of adequate food supply to ensure that the culture animals attain the desired harvesting size within the targeted time. Feeds are among that largest operational cost of shrimp farming and every effort should be made to ensure efficient utilization of feed for growth.

Traditional shrimp farms is most Asian countries employ extensive culture operation in which growth of shrimps fully depends on natural food organisms. In semi–intensive culture, supplementary feeds are given while natural food organisms remain the major food source. Intensive culture operation, shrimp growth is completely depends on artificial diets.

4.7.1 Feeding Behaviour

In the natural habitat, shrimps feed on other small crustaceans, finish, molluses, polychaetes and other slow moving benthic organism. The catch food with their pereopods and take through buccal cavity. They are omnivorous but cannibalize if food is insufficient or of poor quality.

4.7.2 Natural Food

Natural food organisms are allowed to grow in well prepared pond fertilized with organic or inorganic fertilizers. The food organisms in the form of benthic blue–green algae, diatoms, green algae and various species of microscopic zooplankton and microbenthos serve as the natural food (*Lablab, Lumut, Hajas graneinen, Ruppia maritima*) for culture tiger shrimp.

4.7.3 Supplemental Feed

Shrimp farmers provide supplementary feed when natural

food in the pond becomes insufficient. The following types of supplement feed are used:

a) *Moist/Wet Feed* : These are fresh prepared feed using locally available ingredients. The feed should be given immediately after preparation. The commonly used feeds are rice bran with trash fish, chopped toads and frogs, chicken meat, snail meat, mussel meat etc.

b) *Dry Peletted Feed* : Peletted feeds are available commercially. These are also prepared using locally available ingredients usually, commercial pellets are brought in bulk and be property stored in storage room with low humidity to minimize fungi occurrence and insect infection.

4.7.4 Feeding Method

Supplemental feed may be given by broadcasting through feeding tray of automatic machine feeder. Broadcasting method is carried out by spreading the feeds evenly into the pond surface.

Feeding tray containing the feeds are placed strategically at difference parts of the pond. The tray varies in size from 1–10 m^2. The common materials used are bamboo strips or polythene screen.

Various automatic feeders designed primarily for fish were tried in shrimp ponds adopting intensive farming operation where certain amount of feed is given at interval during a 24 hour period.

4.7.5 Feeding Rate and Frequency

Optimal feeding rate and frequency are essential in maximizing conversion rate of feed of shrimp. The more common methods used to determine the feeding rate is:

a) Adjustment of feeding rate through visual observation of left over feed that is employed by the use of feeding tray.

b) Periodic determination of the stock density for appropriate feed ration, which is done by using cast net.

c) Feed ration based on assumed density, which is method of computing feed ration, is based on the estimated survival rate.

The common feeding frequency adopted is 2–5 times a day. Most culturists feed their stock every morning and afternoon only. If the stocks are to be feed 5 times a day, two should be given in day time and 3 at night as the shrimps are move active when dark.

4.8 Harvesting and Packing

The harvesting of semi–intensive ponds is done mostly through drainage Around 80–90 per cent of the stock is removed by this method and rest is collected by hand picking. The harvested shrimp should be washed and placed immediately in chilled water (5–15^{0}C) for about 15 minutes. They are then packed in styrofoam box with alternative layers of crushed ice at a ration of 1:1. Smaller styrofoam boxes are preferred for transportation. Shrimp being fresh is better for I.Q.F. and accelerated freeze–dired packing, which fetch a premium price.

Chapter 4

Hatchery Technology of Giant Freshwater Prawn, *Macrobrachium rosenbergii*

1. Introduction

Early attempts to raise prawns by stocking ponds with juvenile prawns collected from habitats has disappointing results. For a number of years, experiments in Several Asian countries on the rearing of prawn larvae to juveniles under controlled conditions were unsuccessful. In 1961, a major breakthrough was made at the Marine Fisheries Research Institute, Penang, Malaysia, when it was discovered that a certain amount of salinity was an important basic requirement for *Macrobrachium rosenbergii* larvae to survive and develop through their early life stages. News of success spread rapidly worldwide interest in fresh water prawn was generated. Research and development project started in practically all the Asian and Far Eastern Countries, several territories in Ocenia and Micronesia, Several North and South American countries, and some countries in Africa, Europe and in Middle East.

Now this prawn has become the main target species in India. But still now farmers are fully depending on wild fry stocking in farms. Non availability or prawn seed in time is the main handicap of prawn farming activities. Hence the need of the hour is to establish more and more hatcheries for producing seed to meet the increasing demand of the fresh water prawn farmer. Therefore a sustainable hatchery technology for the environmental condition of India may fulfill the increasing demand of the seed of this prawn which enhances the production of prawn (Chakraborty, 1998).

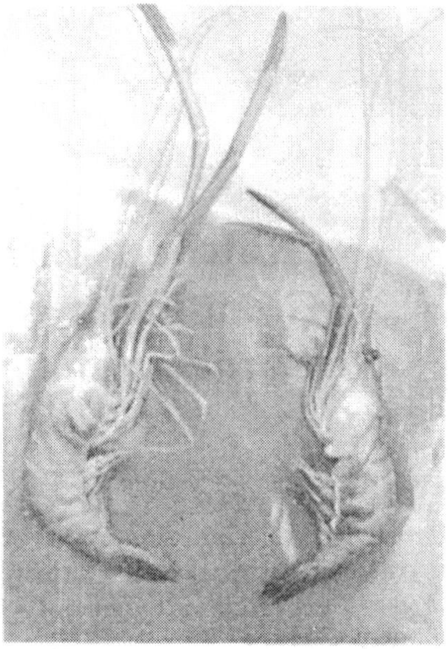

Photo 7: Adult (mature) male and female (with eggs) of *Macrobrachium rosenbergii*.

Photo 8: Different developmental stages of eggs and their colour changes of *M. rosenbergii*.

Photo 9 : Larval rearing tanks (round shape) of *Macrobrachium rosenbergii.*

Photo 10 : *Artemia* hatching in a bachyard hatchery.

Photo 11: Different kits of a hatchery.

2. Site Selection

Selection of suitable site is one the most important factor for success of any prawn hatchery complex. However, for setting up a giant fresh water prawn hatchery, the following factors should be taken into consideration:

a) Adequate supply of fresh water free from pollutants.

b) Adequate supply of salt water, so that the required 12+2 ppt brackish water could be prepared.

c) Easy availability of spawners.

d) Proper arrangement for infrastructural facilities such as – proper arrangement for electricity, accessibility to road etc.

3. Hatchery Design and Construction

The success of a prawn hatchery mainly depends on the appropriate design and establishment of hatchery. For a hatchery complex brood stock development tank (BST), larval rearing tank (LRT), post larval rearing tank (PLT), *Artemia* hatching tank (AT), water storage and mixing tank, over head tank, spawning tank, sand filter, power generation set, biofilter, sea water intake system (if the hatchery is nearer to the sea), air blower or oil free compressor, water and air distribution system, electrical and diesel pumpsets and required. The main hatchery building/shed contain BST, LRT, AT, PLT. The main building should have large windows for proper ventilation. The roof should be covered with corrugated asbestos cements (ACC) and transparent sheet. For proper penetration of light, roof of this shed is made by the ACC and transparent sheet with a ration 3:2 (Chakraborty & Challaya, 1999). Most important parts of the hatchery are sited below:

3.1 Sea Water Storage and Mixing Unit

The necessary brackish water is pumped from sea into hatchery through seawater intake system or by tracking brine or rock salt to the hatchery site and appropriately diluting with fresh water in inland hatchery. This unit includes 3 numbers of tanks (8m × 5m × 1.5m) of which one for brackish/sea storage tank other two respectively. These tanks are of R.C.C.

3.2 Brood Stock Development Tank

The brood stock development tank is generally round or rectangular made with R.C.C. A 50 mm diameter PVC pipe regulates the water flow.

3.3 Larval Rearing Unit

The number of the larval rearing tank depends on the capacity of the hatchery. In medium hatchery RCC tanks are used as larval rearing tank. Dark blue apoxy resin painting is very necessary in the inside of the tank. (Chakraborty, *et al.*, 1998) Round shaped tank is more preferable though rectangular rank is also good. (Chakraborty, *et al.*, 2000). Circular tank of 2 m dia and 1.2 m height is necessary. The slope is maintained towards the centre where out let PVC pipe is attached. Rectangular R.C.C. tank of 4m × 2m × 1.5m height is generally preferred for larval rearing. For proper drainage, L shaped PVC pipe is attached and 1:300 slope is maintained towards the pipe (out let).

3.4 *Artemia* Hatching Tank

Cylindroconical tanks of 0.3 m dia x 0.6 m height and 0.15 m depth slope towards centre having bottom drain facility with a side glass window serve as hatching tanks for *Artemia* cyst. These tanks are mounted on a suitable stand made of iron, which should be coated with fibre glass layer to avoid rusting.

3.5 Post Larval Rearing Tank

The capacity of R.C.C. PL tanks varies from 10 to 30 tons of water. The inner surface of the tank is coated with good apoxy resign or fibre coating. These tanks are rectangular (5m × 2m × 2.5m) with flat or parabolic U shape bottom.

Now a days packing unit is an open shed. The used water from BST, LRT, PLT drained through central of main drain, which ultimately led to outside drain through treatment unit.

Major equipment of giant prawn hatchery include 2 oil free root type air blowers with PVC aeration grid, polythene aeration tubes, air diffusion stones, air regulators and different types of PVC joints, 3 electrically operated water pumps (these pumps should be made of open type gun metal impeller with rust proof centre shaft of stainless steel) and one diesel operated pump with on 5 KVA generator as stand by to electricity.

4. Maintenance and Selection of Brooders

Successful run or a prawn hatchery largely depends upon continuous supply of health broods. Two types of brooders are used – wild and farm. The brooders, which are collected from farm are called *farm brooders*. The brooder, which are collected from river or nature, are called *wild brooders*. Though nature is a good source of brooders but it is not so dependable because of their uncertainties. It is therefore advisable to have the facility for rearing the brood stock of giant prawns as an in built component in the hatchery complex itself.

4.1 Brooders from Brood Stock Pond

It is essential to have brood stock ponds located in the proximity of the hatchery so as to avoid rough handling and long distance transportation of brooders. An impervious clayey soil free from vertical and lateral seepage, having high water retentivity is a desirable criteria in the selection of site. The site of brood stock pond should have perennial source of fresh water, free from pollutants like pesticides etc. While making the layout plan brood stock pond, the following points are looked into:

a) The layout plan of the ponds should be in such a way that they have facilities for easy transfer of prawns from one pond to another from ponds to the hatchery complex without involving much handling.

b) Every pond should have one outlet and one inlet, connected to the supply canal and drain canal respectively.

c) Rectangular ponds of 0.1 ha (80 m × 12.5 m) are ideal for easy maintenance. The number of ponds depends on the production capacity of the hatchery.

The steps of the brood stock pond preparation are enumerated below:

a) Drain and fry the pond after every crop.

b) Repair the bunds, if required.

c) Plough the pond bottom soil in order to expose the sub soil to the atmosphere.

d) After complete drying, analyze the soil pH apply the time as per chart given below:

Soil pH	Dose of Calcium Carbonate (kg/ha)
4.0–5.0	2000
5.1–6.5	1000
6.6–7.5	500
7.6–8.5	200
8.6–9.5	Nil

e) Apply cowdung @ 2000 kg/ha.

f) Fill the pond with fresh water upto 1.0 m depth.

g) Place bamboo or coconut leaves at the bottom for the freshly molted prawns.

The prawn juveniles (PL) of uniform size groups are stoked at the rate of 40,000 to 50,000/ha (4–5 hrs/sqm). Before releasing into pond prawn post larvae should be acclimatized gradually to the temperature and pH of the pond water. The temperature should be adjusted by floating the transport bags in the pond 15 minutes before emptying them into pond water. Stocking should be done in early morning or late evening hours, when the temperature and light intensity are low to avoid stress. Now the prawn should grow for a period of 5 months so as to make their weight 50 to 75 gm. It is recommended that after five months of rearing in grow out pond, adult prawn should be stocked in brood stock development pond with a ratio of one male to three to four female. So that female prawns should be properly utilized (Chakraborty and Chatterjee, 1988). A wide variety of feeds are used for maintaining the brood stock such as rotten rice, oil cakes, small shrimp, foot tissue of apple sail, mussel cut pieces, fish or prawn meal and formulated feed. Feeding should be done 3 time a day. The quantity of feed is adjusted after verifying the consumption after every day. The regular observation of the feeding trays helps the operator in estimating the exact feeding rate water quality, health of prawn etc.

Parameters of water quality play a vital role. The water temperature plays an important role in the growth and survival. The optimum temperature requirement of brood stock pond is from 29^0 to 31^0C. The dissolve oxygen (DO) is one the most critical parameters which controls metabolic activity, growth etc. The minimum DO level should be maintained always 4 ppm. It is better to maintain the DO at saturation level as far as possible. Aeration through pH requirement of brood stock pond is from 7.0 to 8.5. If the pH is high water exchange is carried out, if it is low, lime is added in the pond. The optimum requirement to total alkalinity in brood stock pond ranged from 50 to 100 ppm. This range can be maintained by controlling pH. The algal bloom in brood stock pond is monitored with the help of a secchi disc. A transparency of 50 to 60 cm of greenish brown or light brown colour is good for fresh water prawn. Excessive concentration of algae in brood – stock pond, however, is not desirable. Therefore concentration of algae should be monitored 2 to 3 times daily. If required, the blooms can be controlled by applications of algicides such as clorosan or simazime @ 0.002 ppm.

Periodic sampling should be undertaken by using a cast net at a frequency of once a fortnight. This helps not only in determining the average weight, estimation of population, biomass and growth rate but also assessing the signs of disease, stress, per centage of male and females, feed conversion rate and its efficiency. Large well pigmented and active berried brooders with orange or dark colour eggs are selected.

4.2 Brooders from River or Nature

Wild brooders are collected through various nets mainly dol net. Only a single brooders in a plastic bag is transported at a time. But it is advisable that the brooder prawn should be kept inside the perforated PVC pipe (50–70 mm dia) for transportation.

After insertion of prawn inside the tube both sides of the pipe are closed with 40 mm mesh nylon net to prevent the escaping of prawn. The method of packing not only prevents the spoilage of polythene bag by the strong chelate legs and the rostrum of the prawn brooders during transport but also increases their carrying capacity per unit volume of water inside the polythene bag. Transportation should be made during night time.

5. Larval Rearing

The techniques of larval rearing of this prawn are broadly divided into two system 'green water system' and 'clear water system's. In the "green water system", a bloom of mixed phytoplanton (mainly *Chlorella sp.*) is maintained in the larval rearing tank. The phytoplanton is culture in the saline larval rearing water. The green water is used as replacement water during daily exchange of water instead of plain mized water. The 'green water' culture of more than 25 days old are not used to avoid any bacterial contamination. This system helps is maintenance of good water quality, specially pH and ammonia levels.

The "clear water" system of larval rearing is widely used in most of the hatcheries specially in India because the larvae can be required in higher densities and avoids the requirement of having separate phytoplankton culture under the clear water technique there is the "open systems" as well as the "closed system". In the "open system" more than half of the larval rearing medium is replace daily with fresh and clean water. In "closed system" the remaining medium is recirculated through a sand graval filter or biofilter and allow to pass through the filter bed. The filter water passes through the biofilter at least 10 times a day. To prevent escape of larvae, the suction end is covered with a piece of blotting silk. But 'semiclosed system' is recommended for fresh water hatchery. In 'semiclosed system' higher recirculation of water through a biofilter is arranged as well as a part of a rearing medium is replaced daily according to the availability of fresh 12–2 ppt of water (Chakraborty *et al.*, 1999).

5.1 Larval Rearing Technique

Gravid mother prawns with dark eggs are released into spawning tank where they spawn for four days. After spawning, mother prawns are separated and larvae are placed into larval rearing tank. Now the recommended density of larvae is 6–100/

L. The larvae passes different zoeal stages through metamorphosis. After 27–32 days, larvae are changed into post larvae.

5.2 Feeding of Larvae

The feeding of the larvae is done 3–4 times a day. The feeding starts on second day with the brine shrimp (*Artemia*) nauplii. Up to 10 days, *Artemia* must be needed because of the presence of HUFA (Highly unsaturated fatty acid). After 10 days, chopped *Tubifex* sp., *Bellamya bemngalensis*, prepared feed may be used replacing *Artemia*. The requirement of *Artemia* nauplii for a day should be estimated well in advance depending on the hatching period and the *Artemia* cysts should be kept accordingly for hatching. After hatching, *Artemia* nauplii are harvested by taking advantage of their phototactic nature. For doing so, light is beamed at one place and then the whole cluster is siphoned and filtered on 250 micron blotting cloth. The *Artemia* nauplii so obtained are thoroughly washed and stocked in a storage tank having aeration facilities. Then the nauplii density is estimated randomly. This is done by counting the number of *Artemia* nauplii present per ml and then computing number per litre. The volume of *Artemia* nauplii concentrate to be added to larval tank can be estimated by using the formula:

$$AC = \frac{(DD - RD) \times TV}{AD}$$

Where, AC = *Artemia* nauplii concentration to be added to the larval rearing tank (in lit)

DD = Desired density of nauplii/ml.

RD = Residual density of nauplii/ml.

TV = Larval tank volume (in lit)

Ad = Density of *Artemia* nauplii concentration (nauplii/ml)

when *Artemia* nauplii are fed to the prawn larvae, care should be taken to prohibit the entry or unhatched cysts or shells of cysts. They may cause bacterial infection to the larvae.

Chopped *Tubifex*, fresh water snail *B. bengalensis* or mussel cut pieces should be provided after 10 days replacing *Artemia*. Prepared food may be provided at that time prepared feed consist of fresh water snail/mussel flesh, hens egg, milk powder and vitamin mineral mix or with egg, dry yeast (@1 kg/egg) and vitamin mineral mix. The ingredients are blended in a mixer grinder, coagulated by gentle cooking and particulate by passing through superimposed standard test seives of appropriate mesh size upto stage IV, particle size 200 micron, from stage V to VIII particle size 400 micron and from stage IX to XI particle size 600 micron are used. It is evident that feed should be used in rational ration underfeeding leads to starvation, cannibalism and slow growth. Over feeding (mainly prepared feed) cause pollution. So it is advisable that the bottom of the tanks are cleaned daily in the evening by siphoning out the uneaten food particles and other wastes. Water is to be exchanged to the extent possible with fresh and clean 12 ppt saline water (Chakraborty and Chatterjee, 1998).

5.3 Water Quality Management

Success of a prawn hatchery production on prawn seed largely depends on the water quality management. Sometime synthetic seawater may be used instead of saline water, which is prepared by addition of certain chemicals and common salts in solution. Generally saline water is pumped from the sea. Which is mixed with fresh water for desired salinity (12+2 ppt). Prepared 12 per cent saline water if unfiltered and turbid, may be allowed to settle in a setting tank for 3 days and supernatant water pumped into treatment tank where it is treated with sodium hypo–chlorite and allowed to precipitate for 5 days. To this, 10 ppm sodium thiosulphate is added and aerated vigorously for a day to remove the residual chlorine. The precipitate is then allowed to settle for one day and the supernatant solution is pumped to larval rearing tanks or into overhead tank from where is supplied into various tanks through gravity.

High pH problem sometimes accounted in hatcheries can be corrected by mixing with calculated amount of HCI (or NaOH) applied in divided doses. For maintaining the optimum oxygen level, aeration is carried out almost continuously. Some time temperature goes down (below 24^{0}C mainly in rain season) in such situation, thermostatic heaters are used in the larval rearing tank to maintain the desired water temperature.

5.4 Bio–filtration

Water exchange is a good method of management of water quality in the larval tanks; however, it may be costly in certain situations like inland backyard hatcheries. Biological filters are used as treatment units in such system.

It consists of inert material like pebbles, sand or granite chips mixed with other materials like carbon, lime stone, clam shells etc. filled a container and water is passed through this filter material. Within a few days a film of microorganisms would develop over the surface of the filter material. Because of vigorous microbial activity, the organic content of incoming water is decomposed into simple and less harmful materials. An oxidation reaction is taking place in biological filters. Ammonia, nitrites and other oxides of nitrogen are oxidized to stable nitrates, and sulphides are oxidized to sulphates.

There is a chance for dropping of pH of the effluent water coming from biological filter. In such cases, alkaline calcareous substances lime stone, bivalve shells, corals etc. are incorporated in the filter for the purpose of correcting pH.

5.5 Aeration

Aeration carried out almost continuously helps to maintain sufficient oxygen for respiration as well as for keeping the food particles in suspension. Air passes through water that would drive away the foul gases such as ammonia, hydrogen sulphide etc. formed as metabolic wastes from water.

On an average, larvae are changed into post larvae by 27 to 32 days, which are settled down in the bottom of the tank.

6. Post Larval Rearing

When most of the larvae are settled, the salinity is gradually lowered to fresh water over a period of days usually the remaining larvae also settle by that time. It will be helpful if artificial substrata in the form of seasoned twigs, stiff PVC sesh etc. are provided for early settlement of larvae. An average survival of 20 per cent or production of 20 PL/L could be expected, eventhough it is possible to obtain over 75 per cent survival in individual cases. The chopped feed or prepared feed may be provided as the feed for the post larvae, 3 to 4 times daily.

7. Seed Packing and Transportation

Proper packing is very necessary for transportation of seeds.

The prawn seed (post larvae) are transported to short distance for a duration of 4 to 5 hours in a open container without oxygen. However for long distance of transportation of seeds (upto 36 hrs of journey) oxygenated plastic bags are used.

7.1 Packing of Seeds

Polythene bags of 15–20 lits capacity are used for packing of prawn seed. The corners of the plastic bags are rounded off with the help of rubber bans so as to prevent the prawn seed from getting trapped during packing. For filling the plastic bags, the water from the post larval tank is used. The use of new water is avoided for packing. If new water is used, most of the post larvae will moult during transportation with consequent losses through cannibalism. In order to provide shelter to post–larvae, a few 10 mm plastic straw or nylon thread should be placed in the bags. The avoid cannibalism of post larvae during long distance transportation post–larvae are fed with like feed such as *Artemia* nauplii, before putting them inside the bag. After feeding them properly, the post larvae in appropriate densities (75/L–125L) depending on the distance of transportation. (Chakraborty *et al.* 1998)

After placing post–larvae, the air present in the bag is removed by folding it and remaining 2/3 of the bag is filled with oxygen by inserting the oxygen supply pipe in the water from oxygen cylinder. This helps in saturation of dissolved oxygen in the water. The bag is then sealed tightly with a rubber band. (Chakraborty *et al.* 1998)

For very long distance, the polythene bags are placed in a insulated containers.

7.2 Transportation of Seeds

Normally, if the distance is upto 3–4 hrs. from hatchery to growout pond, they are carried out in open containers by van rickshaw. For long distance, the prawn seed are carried by trucks or trains in India. if facilities are available it is also carried by air.

8. Common Diseases Associated with Larvae Culture

The main consideration in the hatchery phase of this prawn is the health management. The diseases appear in the hatchery due to poor water quality, infected food, over feeding, improper cleaning of the larval rearing tanks etc. Prawn larvae are reared in high densities involving intensive feeding, resulting in heavy loading of the water with dissolve organic matter. Unless the larval rearing is carefully managed, this organic loading can lead to poor water quality and provide excellent substrate for proliferation of primary and secondary pathogens, such as bacteria, fungus and some protozoans.

8.1 Viral Diseases

Incidence of viral diseases have not yet been reported in larvae and post larvae of this fresh water prawn.

8.2 Bacterial Diseases

Reported bacterial diseases are 'black spot disease', clogging of gill, necrosis and development of white patches on tail and base appendages. The 'black spot' diseases is caused by '*Benekea* sp.' filamentous bacteria also sometime encountered in rearing tanks which clogging the gills and interface with respiration. Filamentous bacteria is *Leucothrix* sp. For control of heavy bacterial growth, however, use of penicillin, tetracycline and oxytetracycline are necessary. (Chakraborty & Chatterjee, 1999).

8.3 Fungal Diseases

These infectious are often enhanced due to heavy organic load in the larval rearing medium. If majority of the larvae are infected, it is better to sacrifice the entire batch of larvae before the diseases spread to other tanks in the hatchery. The only remedy is the cleaning of the tank and then disinfect the tanks and other equipments and fill the larval rearing tank with fresh medium. Sometime malachite green and metheline blue are found to control the fungal diseases.

8.4 Protozoan Infection

Protozoan infections are most common in the larvae of *M. rosenbergii*. A number of protozoans including *Epistylis* sp. *Zoothamnium* sp. *Vorticella* sp. and *Cothurnia* sp. may bloom under unhealthy conditions. They attach to the till and exoskeleton of larvae, restricting respiration and mobility. Heavy infections can ultimately result in mass mortality as the larvae become unable to feed, respire or molt. Application of copper sulphate are found to control the protozoal infection.

Diseases are not considered a major problem in freshwater prawn hatchery. However, one should take necessary steps for the prevention of this problem. Proper maintenance of water quality and take hygiene coupled with utilization of proper feed and feeding regime will reduce the incidence of diseases problem. It is advisable to conduct regular examination of different body parts of larvae and their general pigmentation. It is also required to observe the swimming behaviour, feeding behaviour, presence of food in the gut of the larvae along with water quality for proper health management.

Chapter 5

Culture Technology of *Macrobrachium rosenbergii*

1. Introduction

The giant freshwater prawn (*M. rosenbergii)* has been identified as the main target species for farming due to its less cannibalistic behaviour, shorter larval period, fast growth rate, tolerance to a wider range of temperature and high export market value. The commercial culture of this prawn culture have been practiced in India and Bangladesh resulted in very low production. These are the traditional trapping-holding-growing operations, in which young prawns from wild stock are introduced into ponds, tanks, paddy fields and 'bheries' and held in these enclosures to feed and grow to edible size.

In 1961, the techniques for rearing new hatching through all their larval stages were developed by Marine Fisheries Research Institute, Penary, Malaysia. Within a short period the commercial hatcheries assured supply of quality prawn seed to prawn farmer which is a major constraint of its culture. In India, the Prawn Breeding Unit of Central Inland Fisheries Research Institutes developed and indigenous technology in the year 1975.

2. Site Selection

The following criteria must be considered for culturing growout phase in pond.

a) Climate Condition : Warm climate is the most important factor affecting prawn production. Tropical and subtropical climates are most suitable where the culture period is long, while in warm temperate climate, cultivation is possible for a short period only.

b) Topography of the Area : The ideal site should preferable gently slopes i.e. not more than two per cent and pond can be gravity filled and drained.

c) Soil Type : Soil texture should be sandy–clay loam or silty clay loam with 85 per cent water retention. The clay content should not exceed 60 per cent.

d) Water Resources : There are two important sources of fresh water: ground water and surface water. It should be available through out the year, allowing for a ten per cent daily renewal rate of water in all ponds.

3. Design and Construction of Farms

The first step for designing of a giant freshwater prawn is to study thoroughly the survey reports, concerning the soil characteristics, tropography, sources of water supply etc. The design should suit the operational convenience of management and biological necessities of the prawn.

3.1 Size and Depth

For operational convenience and productivity, one hectare pond is widely accepted size. The size may vary from 1/2 hectare to 5 hectare depending upon the local condition. Though square is more economical but rectangular shape is preferred. The constructed pond should hold 1.5 to 2 metre water inside.

3.2 The Dyke

The dyke must be stable and water tight. While designing a dyke the stability should be checked by drawing the hydraulic gradient and there should be a clear cover of at least 0.3 m between the hydraulic gradient line and the toe of the dyke. The side slopes of a dyke primarily depends on the stability of the fill material. The greater the stability of the fill material means steeper the slopes. The recommended side slope of dykes for different type of soils are follows:

Good clay	1 in 1.5 (1 vert : 1.5 hor)
Sandy clay	1 in 2 (1 vert : 2 hor)
Sandy loam	1 in 2.5 (1 vert : 2.5 hor)
Loose and wet earth	1 in 3 (1 vert : 3 hor)

A minimum of 2 m top width should be kept as matte of convenience. If the top of the dyke is used for roadway, the top width should be at least 5m.

3.3 Water Inlet and Drainage

All ponds should preferably be provided with a water inlet. A good water inlet should provide the required and controllable water supply into the pond and should prevent entry of undesirable fish in the pond. We may get an additional advantage of aeration from inlet by putting a bag type screen with the pipe.

The drainage sluice and inlet canal should be at the opposite ends. This helps maximum water movement at the time of water replacement. For drainage sluice, it is safer to have R.C.C., construction sluice should have provision to insert screens of different mesh sizes. It should have provision for lockup also. The sluice has to be operated from the top of the bundh.

The arrangement of ponds should be in such a way that the drainage from all the ponds should pass through a common canal and the inflow should be distributed from a common canal. The main drainage canal and inflow canal should be protected against the entry of unwanted species.

4. Culture Operation

Culture of fresh water giant prawns is a recent development. Early attempts to raise prawns by stocking ponds with juvenile prawns collected from their natural habitats has disappointing results. For a number of years, experiments in several Asian countries on the rearing of prawn larvae to juveniles under controlled conditions were unsuccessful. After successful rearing in sufficient number of juveniles, initiate in culturing juveniles to large prawns widely.

In India, the three major types of prawn culture systems are seen in three ways – traditional or extensive culture, semi–intensive culture and intensive culture.

4.1 Extensive Culture

The traditional of extensive method has been in practice for many years especially in Bangladesh and India, mostly in West Bengal, Kerala, Karnataka and Goa. In this method, young prawns (from wild stock) are collected separately and stock into ponds, tanks, paddy fields and 'bhrics' and held in these enclosures for growing to edible size. Or, in natural way, the prawn seed entering with the tide are trapped and harvested after 5–6 months. There is no feeding of fertilization involved. Prawns largely depend on the natural food. Production from such system are less than 500kg/ha/yr.

In Kerala trapping the juvenile prawns, rearing them to bigger size and harvesting them is done in paddy fields adjacent to the backwater areas. This method is popularly termed paddy cum prawn culture. Usually, *Penaeus* and *Metapenaeus* are the more prevalent species of prawns whereas the *Macrobrachium* production is minimum.

Similar way, in West Bengal the traditional prawn culture is known as Bhasabadha practiced in brackish water area of Sunderban. The most important prawn obtained is *P. semisulcatus* and others are *P. monodon* and *Palaemon styliferous*.

Advantages

a) Little expenditure is involved in stocking the ponds.

b) Maintenance of the ponds is not very expensive.

c) Artificial feeding is not necessary.

d) Expensive equipments are not necessary.

e) Harvesting can be done without much extra expenditure.

Disadvantage

a) Stocking is indiscriminate — neither the species nor the number stocked can be controlled.

b) Sticking procedure allows entry of predators and undesirable species.

c) Productivity is low and the yield fluctuates widely.

d) Total harvesting is not possible as it is difficult to drain the ponds completely.

4.2 Semi–Intensive Culture

In semi–intensive system for improved culture the disadvantages of the traditional practices may be eliminated by selective stocking, eradicating weeds and predators, liming, fertilization, fed with natural and artificial food throughout the cultural period. The culture period varies between 4–6 months. The rate of stocking depends on whether the prawns are to be cultured alone (monoculture) or in combination with suitable species of fish. In monoculture, a stocking density of 15,000–30,000/ha is desirable, depending on the age and condition of the pond. In polyculture, the stocking density is reduced to 50 per cent compatible fish fry viz. catla, rohu, silver carp, grasscarp etc., are stocked before stocking the prawn seed. Bottom feeders viz. common carp, mrigal, calbasu etc., are not a good combination. In India, yields ranging from as low as 284 kg to as high as 1,640 kg/ha in 4 months have been recorded in monoculture and from 85–709 kg/ha in polyculture in 6 months. This method is widely practiced in India and Bangladesh. The details descriptions of this method has been included later on.

Advantages

a) Moderate feed cost.

b) Expensive equipment may be avoided.

c) Maintenance cost of pond is minimum.

d) A little chance of disease infestation.

Disadvantages

a) Productivity is moderate.

4.3 Intensive Culture

The intensive system has been adopted by medium and large–scale farmers and it involves the provision of good quality supplementary feed, higher stocking densities, regular exchange of water and aeration. In intensive system, proper management also required for hatchery phase, which covers the management of broodstock, incubation, larval rearing and seed acclimatization to fresh water.

Advantages

a) Productivity is very high.

Disadvantages

a) Very high capital investment required.

b) High density culture is vulnerable to diseases.

c) Chances of high mortality if the mechanical failures occurs for a short time.

d) Supervision required for 24 hours.

5. Transportation of Seeds to the Farm Site

Post larvae should be transported to the pond in the early morning or late evening to avoid subjecting them to heat stress. Cooled and aerated prawn transport tank would be the ideal for transporting prawn larvae but they are rarely available.

For shortest distance (upto 1 hour) post larvae can be transported in cans, plastic tanks or packed in plastic tanks or packed in plastic bags. For transport in cans or tanks of 100 lit capacity may be used holding 40 lit of water with 30,000 post larvae. A few weed or plastic strips are added to the packet to prevent excessive water movement, shelter and feed to prevent cannibalism.

In case of long distance transport, the seed are transported in plastic bags supplied with oxygen. A 45 x 80 cm bag holding 8 lit of water will take 1,000–2,000 post larvae, normally the number of seed 125–250 post larvae/lit is used, though it may vary depending upon distance and temperature.

Firstly the required number of post larvae are placed in double–layered plastic bag containing 1/3 water 2/3 air. Oxygen is supplied to the bags after which they are tied with rubber bands. The bags are then placed in "Styrofoam" or cardboard boxes, which can be used to ship post larvae by air most effectively. Enclosed in non–insulated boxes can be used for night (cool) journeys by rail. For long day–time (hot day) journeys, the transport box is insulated and a plastic bag of ice is placed within the box to keep the temperature cool which reduces metabolic activity during transport. Mortality during transport is caused by injury to the soft seed (moulted) and cannibalism. It's the inter–moult period at this stage is very short no remedial measure is possible.

6. Water Quality Management

The water quality parameters with their optimum ranges for successful prawn culture are given below:

Water temperature	:	28^0 x 31^{0}C
pH	:	7.0–8.5
Dissolved Oxygen	:	above 3 ppm
Total hardness	:	40–100 ppm
Calcium	:	less than 100 ppm
Phosphorus	:	less than 1 ppm
Nitrates	:	less than 1 ppm

6.1 Water Temperature

Temperature has a crucial role on growth rate, since prawns are unable to regulate their body temperatures and solubility of DO in the water (9.2 mg/litre at 20^{0}C and 7.7 mg/litre at 30^{0}C).

Regulation of temperature is a difficult process in the pond. The pond culture period must be done when the temperature synchronize with the particular season.

6.2 Dissolved Oxygen

In aquaculture practice, DO is one of the most critical parameters. In natural environment, there is normally sufficient oxygen in water to support aquatic organism. But under aquaculture practice, we often find the demand of oxygen by the prawn and other organisms are extremely high because the culture organism are stocked at high densities and microbial breakdown organic matter.

It is normally towards the end of night the maximum depletion in oxygen content take place. Normally prawns leave the bottom and go up to the edge of the pond. In case of prawn, the optimum range of DO concentration is between 5–7 mg/litre, which is required for normal growth. At 2 mg/litre oxygen, concentration, the prawns become lethargic and found to crawl out of the pond margin, their muscles become opaque and they die. To overcome this problem, a quick remedy is mixing and aerating the pond by mechanical means through paddles to stir up the surface of the water or air injector. Alternatively, quick release of water into the pond through pump or admitting feeble flows of water into the pond after midnight such phenomenon can be averted.

6.3 pH

pH plays an crucial role on chemical reaction of the pond and in the physiological activity of prawns (exchange of ions across cell membrane). It increases as a result of oxygen produced by photosynthesis and it decreases at night due to generation of CO_2 from respiration of all living organisms in the pond. Acidic water in ponds can be naturalized. The most common treatment for acidic water is the addition of Agricultural lime calcium carbonate ($CaCO_3$), quicklime (CaO), dolomite [$CaMg(CO_3)_2$] or builder's lime – calcium hydroxide $Ca(OH)_2$ to the water.

On the other hand, the more the pH increases, the more ammonia, which is excepted by prawns, is in the form of toxic MH_3 rather than ammonium (NH_4^+). At 30°C at pH 7.0 the ammonia is broken down into 0.7 per cent NH_3 and 99.3 per cent NH_4 whereas at pH 9.0 ammonium converted into 40 per cent NH_3 and 60 per cent NH_4. Ammonium sulphate has been suggested as a treatment for high pH.

6.4 Hardness

Hardness originally referred to the ability of water to precipitate soap. The precipitation is largely a function of the concentration of calcium and magnesium in the water, but may also be caused by ions of several metals as well as by hydrogen ions. Hardness has now come to mean the total concentration of calcium and magnesium ions expressed in terms of ppm of calcium carbonate. Soft water is 0–55ppm, slightly hardwater is 56–100ppm, moderately hard is 101–200pp, and very hard water is 201–500ppm.

Calcium is important in culture pond since it is required by prawn for growth and to harden the outer shell. If the water is below the optimum level, it must be provided with extra calcium and magnesium in the pond.

7. Feed Management

Feed management play a vital role in *M. rosenbergii* culture. Natural food is the most important factor for culture. But, due to lack of natural food in modified extensive, semiintensive and intensive culture, supplementary food are always used.

7.1 Natural Food

In pond environment, giant freshwater prawns are able to satisfy their nutrient requirement from a variety of sources, which include aquatic worms, insects, small molluses, crustaceans, plankton and organic detritus. However, under culture condition, natural food is not sufficient due to increasing stocking density. The production of natural food can be enhanced by fertilizing fish ponds with organic manure, inorganic fertilizer or a combination of both. Although natural food items have very good conversion values, but the difficulty to get these in large quantities regularly all through the seasons and their easily perishable nature make supplementary food those suitable for prawn culture. In modern methods of feeding natural food constitute 30 per cent and artificial 70 per cent of the feed given, because natural food is physiologically complete.

7.2 Supplementary Food

In semi–intensive and intensive culture the natural productivity is not sufficient and it is necessary to provide supplementary feeds to increase growth, better survival and production. Before applying any supplementary feeds as must consider the nutritional requirements of the grow out fresh water prawn. The known nutritional requirements of the fresh water prawn is:

Nutrients	***per cent of dry weight***
Protein	25–30
Lipid	4–6
Minerals	12–16
Fiber	10
Vitamins A, D_3 , E, C and B_6	0.5

Supplementary feeds used in semi–intensive culture mostly consists of agricultural by–products or by–products of the animal husbandry. Locally available cheap feeds such as broken rice, tapioca root, trash fish, foot of pila, bivalve meat etc. are used as supplementary feeds, the vegetable and animal feeds separately mixed together in the ration of 1:3 with 20–25 per cent protein content will give better results. Fresh water prawn farms which make their own feeds are in Thailand and other Asian countries, particularly in small scale.

There is very little information on the efficiency of this form of feeding.

The reasons for the decreased ineffectiveness of this form of feeding are manifold, significant wastage of food individual fish being unable to ingest sufficient quantities of each of the constituent ingredients and thereby not obtaining nutritionally favourable diet as well as the digestibility of the individual ingredients. Even though farm made feed can be environmentally more damaging than commercial feed, because of its higher biological oxygen demand (BOD). The improper feeding practices is not only results in unnecessary waste but also leads to more rapid (bottom) pond soil deterioration due to left over uneaten feed, which become rapidly fouled, giving rise to poor water quality. In this way, growing stressful conditions, loss of appetite, poor growth and disease outbreaks causing high mortality can occur.

To overcome this problem the artificial compounded food is more suitable for use in prawn culture. There are several considerations that must be recognized when a diet is being formulated. Not only should the diet be nutritionally sound, but it must also be in a form that is acceptable to the cultured organism.

Some factors that may affect an animals acceptance of a feed are the size, colour, smell and hardness of the pellet. The standard feed should contain sufficient quantities of carbohydrates, proteins, fat, minerals and vitamins to satisfy the known requirements of prawn in terms of both quality and quantity. Some of the quality an quantity. Some of the quality requirements are known and many compound feed are made available in the market according to these requirements.

Table : Examples of Thai Feed Formulae for Fresh Water Prawn Farm Fingerling to 3 Month (30 per cent Protein)

Ingredients	*per cent*	per cent
Formula No 1		
Fish meal	15	–
Shrimp head meal	42	–
Soyabean meal	10	–
Coconut oil cake meal of trash fish	15	–
Wheat flour or tapioca flour or rice bran	13	–
Vitamin and mineral premix	5	–
Formula No 2		
Trash fish	–	35
Fish meal	–	17
Broken rice	–	15
Ingredients	per cent	per cent
Rice bran	–	18
Soyabean meal	–	12
Vitamin and mineral mix	–	3
Total	100	100

Table : Feed Formula for Freshwater Prawns in Thailand 1 Month Before Harvest (25 per cent Protein)

Ingredient	*per cent*
Fish meal	10.0
Shrimp head meal	25.0
Soyabean meal	10.0
Broken rice	25.5
Rice bran	25.5
Vitamin and mineral premix	4.0
Total	100.0

The artificial feeds will have to be produced in pellet from prawns, which are in the habit of holding the food while feeding. The pellets should sink in water quickly at the same time keeping the shape for atleast a few hours till they are consumed completely by the prawns.

7.3 Feeding

Prawns are generally fed once a day only. Pellets may be distributed uniformly over the surface of the pond or placed in wide containers set up at several points along with the margin to the pond during evening hours, since adults are particularly active at night. The rate of feeding depends on the age of the prawns and the temperature and recommended doses are given below:

Table : Calculation of Required Quantity According to Age of the Population

Age (month)	1	2	3	4	5	6	etc.
% of the biomass	8	6	4	3	2	2	2

8. Harvesting and Marketing

Visually pond reared giant prawns reaches marketable size within 5–8 months depending on the stocking rate and other conditions. Normally all the giant prawns do not grown at uniform rate and disparity in size appears in spite of stocking the seed from the same brood. Hence it is desirable to check the optimum size for harvest from time to time.

Two principal harvesting systems are followed, 'partial' and 'complete'. Partial harvesting consists of a system for removing the larger prawns from a pond at regular, often monthly, intervals. This is accomplished with a specially designed net (mainly cast net) (5 cm mesh) which will capture the larger prawns and allow the smaller one to escape. In some areas, a seine with smaller mesh may capture most of the prawns in a pond. The larger prawns are then selected for market and the smaller ones returned to the pond.

In complete harvesting system, prawns are harvested by draining the pond. This is carried out in the early morning to avoid overheating the water when there is only a little water left. At final stage when the prawns are near the outlet, they must be provided with fresh oxygenated water near the outlet, they must be provided with fresh oxygenated water near the outlet channel to attract them. After harvesting, prawns are put into a cage in the pool for the first sorting.

The popularity of the prawns among consumers continues to increase, causing a steady increase in market demand.

The domestic demand for giant prawns is so great that prawn harvested from ponds are quickly absorbed by local markets. Ordinarily, prawns are harvested, and sent to the market early in the morning, and sold within a few hours. Most of them are kept fresh under ice flakes, some are kept alive in tanks or cisterns with running water, or in shallow bamboo baskets covered with fine aquatic plants and kept moist and cool by frequent sprinkling with ice cold water.

Apart from these local markets, exports to the principal sea prawn consumers countries, USA, Japan, and Europe come from the large natural resources of countries in the south east Asia.

Chapter 6

Economics of Hatchery and Culture

1. Economics of a *Macrobrachium rosenbergii* Hatchery

For the establishment of freshwater hatchery of over 10 million seed production capacity, the investment required will be of the order Rs. 23 lakhs. About 80 per cent of the total cost are required for generating fixed assets. Cost of land and its improvement works out 28 per cent of total. Building including hatchery and post larval rearing tanks demand about 25 per cent. Construction cost of brood stock development pond and storage and mixing tank amounts of 19 per cent. Cost of equipment and fittings including hatchery amounts to 26.5 per cent. Profitability of hatchery depends on the quantity of the seed produced, i.e., the soundness of technology it is adopting, the efficiency of the operation of the hatchery and management of different production components. A net profit of 42.63 lakhs is possible at the end of the year of operation after interest and depreciation. The economics of a coastal hatchery with a capacity of 10 million post larvae per year is given below:

A. Capital Investment	
1. Cost of Land – 2 ha	4,50,000.00
2. Civil Works:	
a) Filling of land, compound wall	2,00,000.00
b) Hatchery building with different RCC tanks, lab room, overhead tank	3,50,000.00
c) Office building, packing room, rest room & kitchen etc.	1,75,000.00
d) Generator and blower room	50,000.00
e) Brooder ponds, settling tanks, filter & storage tanks etc.	4,50,000.00

	3. Equipments:	
	a) Diesel generator set	50,000.00
	b) Pump set (5 nos.)	65,000.00
	c) Air blowers (2 nos.)	75,000.00
	d) Deep freezer	25,000.00
	e) Furniture & fitting	25,000.00
	f) Laboratory equipment	1,00,000.00
	g) Filter, UV sterilizer	2,00,000.00
	h) Artemia tanks, sieves etc.	30,000.00
	i) 3 phase electric connection, water supply, telephone	40,000.00
	Total	6;10,000.00
B.	Operation Cost	
	1. Pay and Allowances:	
	a) Manager, 1 no. (5000 x 12)	60,000.00
	b) Technical person, 2 nos. (2 x 4, 500 x 12)	1,08,000.00
	c) Fisherman cum watchman, 6 nos. (6 x 1500 x 12)	1,08,000.00
	2. Cost of Feed	5,00,000.00
	3. Contingencies	2,25,000.00
	Total	10,01,000.00
C.	Interest	
	1. Interest on fixed capital 23 lakhs @ 20 per cent	4,60,000.00
	2. Interest on working capital 10.1 lakhs @ 20 per cent (6 months)	1,01,000.00
	Total	5,61,000.00
D.	Financial Analysis	
	1. Income from 10 million seed @ Rs. 60/100	60,00,000.00
	2. Less: Operation cost	10,01,000.00
	3. Gross profit	49,99,000.00
	4. Less:	
	a) *Interest*	5,61,000.00
	b) Depreciation	1,00,000.00
	c) Insurance	75,000.00
	Total	7,36,000.00
	Total Less	7,36,000.00
	Net Profit	42,63,000.00

2. Economics of Giant Freshwater Prawn Farming

Depending upon existing infrastructural facilities, location of the sit, distance from the water source, topography and various other parameters as well as technology adopted for culture (extensive, semi–intensive or intensive), the development cost as well as operation cost of a M. rosenbergii farm may vary. The economics of a prawn farm (10 ha) is given below:

Assumption

1.	Total area (ha)	10
2.	Water spread culturable area (ha)	8
3.	Size of pond (ha)	05
4.	a) Stocking density (per sq.m.)(Prawn)	10
	b) Catla & Rohu	500 no/ha (Catla)
5.	No . of crop in a year	1,000 no/ha (Rohu)
6.	Expected survival (per cent)	1
7.	Average weight of individual after end of culture period	60
		70 gm
8.	Anticipated production rate (tones/ha/year)	4.2
9.	Expected production (2.7 x 8 ha) of prawn	33.6 ton
10.	Expected production of crop (1500 kg/ha)	12 ton

A Capital Investment

1.	Survey, site clearance of development of infrastructural facilities	1,50,000.00
2.	Earth work for pond dykes construction and excavation of drainage channel	7,50,000.00
3.	Feeder channel	3,00,000.00
4.	Inlets gets (20 nos)	1,00,000.00
5	Outlets (10 nos.)	1,00,000.00
6.	Pumphouse cum stores	1,00,000.00
7.	Office, lab and dormitory staff room	1,50,000.00
8.	Watchman shed	20,000.00
9.	3 No. of 25 HP pump	1,25,000.00
10.	1 NO. 25 HP pump with diesel	75,000.00
11	Electrification	1,50,000.00
12	Laboratory equipments	1,00,000.00
13.	Office furniture	50,000.00
14.	Transport vehicle	2,55,000.00
15	Misc.	80,000.00
	Total	**25,00,000.00**

B.	Working Capital (Recurring Expenditure)	
	1. Pond preparation (Mahua oil cake, lime, cowdung, fertilizers etc.)	1,16,000.00
	2.	
	a) Cost of M. rosenbergii post larvae seed	4,00,000.00
	b) Cost of carp finger lings	6,000.00
	3. Feed (43 tones) (Prawn + carp)	19,00,000.00
	4. Fuel charge and energy charge	1,75,000.00
	5. Office expense	1,20,000.00
	6. Pay & Allowances	
	a) Farm supervisor (Rs. 3000/months)	72,000.00
	b) 1 pump operator (Rs. 1500/month)	36,000.00
	c) 1 store clerk cum cashier (Rs. 1500/month)	18,000.00
	d) 5 watchman (1000/month)	60,000.00
	7. Insurance premium, Harvesting & marketing charge @ Rs. 1/kg)	2,00,000.00
	Total (say)	31,00,000.00
C.	Financial Analysis	
	1. Capital cost	37,00,000.00
	2. Working capital for 1st crop	31,00,000.00
	3. Total loan component	68,00,000.00
	4. Income from 1st crop	
	a) Prawn (@ Rs. 250/kg) (33.6 ton)	7,28,000.00
	b) Carp (12 ton)	2,76,000.00
	Total: (a) + (b)	80,04,000.00
	5. Repayment and recurring cost.	
	a) 1/5 of loan repayment	10,20,000.00
	b) Interest on loan amount @ 16.5% per annum	8,31,500.00
	c) Operational cost for the next crop	31,00,000.00
	d) Depreciation @ 5 per cent on capital cost	1,85,000.00
	Total: (a) + (b) + (c) + (d)	51,46,000.00
	6. Expected net profit at the end of first crop (1st year)	28,58,000.00
	7. Annual net profit from 6 years (after loan is fully repaid i.e. Rs. 4.719 Lakh/ha/year 47.19	47.19

3. Economics of *P. monodon* Hatchery

For the establishment of a P. monodon hatchery of over 30 million capacities, the investment required will be the order Rs.

24 lakhs. Profitability of hatchery depend on the quantity of the seed produced i.e., soundness of technology it is adopting. A net profit of 40.63 lakhs is possible at the end of the year. The economics of a tiger prawn hatchery with a capacity of 30 million post larvae per year is given below:

A.	Capital Investment	
	1. Cost of land – 2 ha	4,50,000.00
	2. Civil Works:	
	a) Filing of land, compound wall	2,00,000.00
	b) Hatchery building with different RCC tanks, lab room, overhead tank	3,50,000.00
	c) Office building, packing room, rest room & kitchen etc.	1,75,000.00
	d) Generator and blower room	50,000.00
	e) (Brooder) ponds, settling tanks, filter & storage tanks etc.	3,50,000.00
	3. Equipment:	
	a) Diesel generator set	50,000.00
	b) Pump set (5 nos.)	65,000.00
	c) Air blowers (2 nos.)	75,000.00
	d) Deep freezer	25,000.00
	e) Furniture & fitting	25,000.00
	f) Laboratory equipment	1,00,000.00
	g) Filter, UV sterilizer	2,00,000.00
	h) Artemia tanks, sieves etc	30,000.00
	i) Out door Algal tank	1,00,000.00
	j) 3 phase electric connection	40,000.00
	Total	6,10,000.00
B.	Operation Cost	
	1 Pay and Allowances	
	a) Manager, 1 no. (5000 x 12)	60,000.00
	b) Technical person. 2 nos (2 x 4. 500 x 12)	1,08,000.00
	c) Fisherman cum watchman, 6 nos (6 x 1500 x 12)	1,08,000.00
	2. Cost of feed	5,00,000.00
	3 Cost of wild brooders	2,25,000.00
	4 Contingencies	2,00,000.00
	Total	12,01,000.00

C. Interest

1. Interest on fixed capital 23 lakhs @ 20 per cent	4,60,000.00
2. Interest on working capital 12.1 lakhs @ 20 per cent (6 months)	12,000.00
Total	5,81,000.00

D. Financial Analysis

1. Income from 30 million seed @ Rs. 20/100	60,00,000.00
2. Less: Operation cost	12,01,000.00
3. Gross profit	47,99,000.00
4. Less:	47,99,000.00
a) Interest	5,81,000.00
b) Depreciation	1,00,000.00
c) Insurance	75,000.00
Total	7,56,000.00
Total Less	7,56,000.00
Net Profit	40,43,000.00

4. Economics of a Semi Intensive Farming of *P. monodon*

A. Capital Investment (Fixed)

1. Cost of land 10 ha	25,00,000.00
2. Civil Works	
a) Survey of site clearance development of infrastructure facilities such as approach road	1,50,000.00
b) Earth work, pond dykes constructions, and excavation of drainage channel	8,50,000.00
c) Construction of feeder channel	3,00,000.00
d) Pond inlet and out lets	2,00,000.00
e) Construction of Pump house & Stores	1,00,000.00
f) Office cum lab and dormitory for staff	1,50,000.00
g) Watchman shed	25,000.00
Total	17,75,000.00
3. Equipment	
a) Diesel generator set	75,000.00
b) Electrification	1,50,000.00
c) Pump set (3 nos.)	75,000.00
d) Laboratory equipments	1.00,000.00
e) Purchase of Transport vehicle	2,50,000.00

	f) Office furniture	50,000.00
	h) Paddle wheel aerator	4,00,000.00
	Total	11,00,000.00
B.	Operation Cost	
	1. Seed @ 20/sqm2 i.e. 2 lakh/ha x 8 hc	3,20,000.00
	2. Feed	5,00,000.00
	3. Fertilizer, lime, cowdung etc.	1,50,000.00
	4. Pay and allowances	4,00,000.00
	5. Contingencies	2,50,000.00
	Total	11,00,000.00
C.	Financial Analysis	
	1. Fixed capital	28,75,000.00
	2. Working capital	16,20,000.00
	3. Total Project cost	69,65,000.00
	4. Total loan component	52,47,000.00
	5. Income from first crop	80,00,000.00
	6. Repayment and recurring cost	
	a) 1/2 of loan repayment	10,49,000.00
	b) Interest of loan @ 18.5 per cent	9,71,000.00
	c) Operation cost of next crop	16,20,000.00
	d) Depreciation @ 5 per cent capital cost	1,44,000.00
	Total	37,84,000.00
	7. Expected net profit at the end of 1st crop	42,16,000.00
	8. Expected net profit from 3rd yr. (after loan is fully repaid)	2,81,69,000.00

Assumption

1. Total area (ha)	10
2. Water spread cultivable are	8
3. Pons size (ha)	0.5
4. Stocking density (Per sq. m.)	20
5. Culture period	120 days
6. Total crop (per year)	2 nos.
7. Survival rate	50 %
8. Average harvesting weight of individual shrimp	75
9. Expected selling price (Rs./kg)	400

References

AQUACOP (1983): Constitution of broodstock, maturation, spawning and hatching system for penacid shrimps in the centre oceanologique du pacifique. In Hand book of Mariwlture. vol. 1. Crustacean Aquaculture (Ed by J. P. McVey) CRC Press. Boca Raton. 105–127.

Chakraborty, C. (1998): Seed production technology of giant fresh water prawn, Macrobrachium rosenbergii (deMan) In: Training manual, Indian Institute of Technology (Agriculture & Food Dept.), Kharagpur: 27–33.

Chakraborty, C., Chatterjee, T. K. (1998a): Studies on feeding pattern of fresh water prawn larvae Macrobrachium rosenbergii (deMan) in hatchery system. Trans. Zoo. Soc. India. 2(2): 24–29.

Chakraborty, C., Chatterjee, T. K. (1998b): Assessment of serration in the brood stock development pond for optimum production of berried female: A critical factor for continuous seed, production in freshwater prawn, Macrobrachium rosenbergii (deMan) J. Aqua. Bio 13 (1) 34–36.

Chakraborty, C., Chatterjee, T. K. and Chakraborty, S. K. (1998): Effect of rearing tank background colour on survivability of giant fresh water prawn larvae, Macrobrachium rosenbergii (deMan). Trans. Zool. Soc. India. 2(1): 42–45.

Chakraborty, C., Chatterjee, T. K. and Chakraborty, S. K. (1998): Development of indigenous larval rearing technology for the seed production of fresh water prawn, Macrobrachium rosenbergii. Proc. International Conference on Ecological Engg.: In press.

Chakraborty, C., Chatterjee, T. K. and Chakraborty, S. K. (1998b): Evaluation of Packing and Transportation technology of fresh water prawn, Macrobrachium rosenbergii (deMan) in West Bengal. Abstract. Nat. Symp. & Workshop on Finfish & Shellfish Farming (diseases and control organised by dept. of zoology, University of Calcutta): 45.

Chakraborty, C., Chatterjee, T. K. and Chakraborty, S. K. (1999): Water quality of larval rearing system of giant fresh water prawn. Env. & Ecol. 17(2) 432–435.

Chakraborty, C., Chatterjee, T. K. (1999a): Effects of stocking density and light intensity on growth & survival in larval rearing of fresh water prawn, Macrobrachium rosenbergii. Environ. & Ecol. 17(2): 288–290.

Chakraborty, C., Chatterjee, T. K. (1999b): Antibiotic resistant Aeromonas hydrofila with R Plasmid DNA from larvae of Macrobrachium rosenbergii: A threat to aquaculture. Proc. Env. Bio.

Fox, J. M. (1983): Intensive algal culture technique. In: Hand Book of Mariculture, vol. 1. Curistacean Aquaculture (Ed. by J. P. Mc Vey) CRC Press Bola Raton 15–41.

Gopalakrishnan, P. (1998): Recent advances in shrimp farming in India. In: Advances in Fisheries and Fish Production. vol. 2. (Ed by S. H. Ahmad). Hindustan Publishing Corporation, New Delhi, India, pp. 142–170.

Gopinathan, C. P. (1995): Micro Algal culture. In: Live food. (Marine Products Export Development Authority Kochi, India) p. 1–16.

Hudinaga, M. and Kittaka, J. (1967): The large scale production of the young kuruma prawn. P. Japonicus Bate. Bull. Plank. Soc. Japan, 13: 35–46.

Hudinaga, M. (1942): Reproduction, development and rearing of Penaeus Japonicus Bate. Japan. J. Zool. 10(2): 305–393.

Kungrankij, P. (1985): Overview of penaeid shrimp culture in Asia. In: Proceedings First International Conference on Culture of penaeid prawns shrimps (Ed. by Y. Taki, J. H. Primakera and J. A. Llobrera). SEAFDEC Aquaculture Department, Philippines: 11–21.

Liao, I. C. (1985): Brief review of the larval rearing Techniques of Penaeid prawns. In: Proceedings First International Conference on Culture of penaeid prawns / shrimps (Ed. by Y. Taki, J. H. Primakera and J. A. Llobrera), Iloilocity, Philippines.

Mock, C. R. and Neal (1974): Penaeid Shrimp hatchery system. Symp. Aquacul. Latin America, CARB/6/74/FAO/CARPAS/SE. 2–9.

New, M. B. and Singholka, S (1985): Fresh water prawn farming. A manual for the culture of *Macrobrachium rosenbergii*. FAO Tech. Pap. Rev. (225) 118 pp.

New, M. B. (1995): Status of fresh water prawn farming: A review. Aquaculture Research, 26, 1–54.

New, M. B. (1988): Fresh water prawn: Status of global aquaculture, 1987. NACA Tech. Manual No. 6. NACA, Bangkok, Thailand, 58 pp.

Pillay, T. V. R. (1993): Shrimps and prawns. In: Aquaculture Principal and Practices. Fishing New Book (division of Blackwell Science Ltd.). Cambridge. 425–459.

Rami Reddy, P. (1995): Penaeid Prawn hatchery management for mass scale seed production. In: Short Term Training Programme on Management of Brackish Water Fish Farm. AT ARTC Kakinada a division of C.I.F.E. Bombay, India 44–50.

Rao, K. J. and Tripathi, S. D. (1993): A manual of giant fresh water prawn hatchery, ICAR Manual Series (2) 50 pp.

Santhanam, R., Ramanathan, N., Venkataramanujam, K. and Jegatheesan, G. (1987): Phytoplankton of Indian Seas. Published by Daya Publishing House New Delhi, India. 134 pp.

Index

A

B

C

D

P

R

S

T

V

W

Z